Finish Carpentry

Efficient Techniques for Custom Interiors

by Gary Katz

The Journal of Light Construction • Craftsman Book Company

Copyright ©2001 by The Journal of Light Construction and Craftsman Book Company.
All rights reserved.
First printing: November 2001   Second printing: May 2002   Third printing: November 2003
Fourth printing: December 2004

Edited by Rick Mastelli
Illustrated by Tim Healey
All photographs by the Author

International Standard Book Number: 1-928580-20-3
Library of Congress Control Number: 2002105779
Printed in the United States of America

A Journal of Light Construction Book in conjunction with Craftsman Book Company.
The Journal of Light Construction is a tradename of Hanley Wood, LLC.

The Journal of Light Construction
186 Allen Brook Lane
Williston, VT 05495

Craftsman Book Company
6058 Corte del Cedro
Carlsbad, CA 92009

To Lawrence, Tristan, Robin and Whitney.
And to my mother, who said I'd never be a very good carpenter.

About the Author

Gary Katz is a licensed general contractor based in southern California. For more than 30 years, Gary has specialized in finish carpentry working on a wide range of projects — from tract housing to expansive custom homes, and from commercial tenant improvement to schools and hospitals.

During that time Gary has trained many carpenters on everything from basic techniques to advanced design and joinery. In addition, Gary is a popular presenter at national trade shows such as JLC LIVE and The Woodworking Shows, where he demonstrates finish carpentry techniques and leads carpentry seminars and hands-on workshops.

For more than a decade Gary has shared his expertise as a frequent contributor to *The Journal of Light Construction* and other leading trade magazines, and he is moderator of JLC's online carpentry forum. He is author of the recently published *The Doorhanger's Handbook*.

Acknowledgements

First, I want to thank Iverson Construction: Mike And Brian Iverson allowed me to photograph our crew install finish work in several of their fine custom homes. Without their patience and their projects, the quantity and quality of the images in this book would not have been available. My appreciation also goes to all of the carpenters I've met — either in person, on the telephone, or online, who have shared with me their tips and techniques as I have shared mine: Kevin Coyner, Jim Bodget, Frank Maglin, Mike Guertin, Sandor Nagyszalanczy, Rob Zschoche, Kirk Grodske, Steve Phipps, and countless others for whom there isn't room here to name.

I especially thank all of the carpenters who have worked with me daily and whom I have learned from: Jim and Mark Cross, Bill and Mike Clark, Eddie Soleri, Adam Guerroro, Larry Rose, and Mike Rutledge.

I thank the Shaefer Brothers, Al and Royal, without whose help I couldn't have written anything about hanging doors.

And I thank Mike Sloggatt, Gary Ashburn, and Jed Dixon for teaching me so much about angles and arches. I thank Jed, too, for sharing his deep interest in architectural woodwork. To Don Dunkley and the entire crew of JLC LIVE (Tom Brewer, Tom Carty, Butch Clark, Carl Hagstrom, Michael Byrne, and Bill Robinson) — my special thanks for inviting me to share the reward of teaching other carpenters. The experience has changed my life.

Then there's the photography: None of the images would have been possible without the inspiration of Dean Della Ventura. Dean has shared with me the secrets of his craft. In the same breath, I thank Roe Osborn for sharing his photographic tips and techniques, and Kirk Grodske for his expert advice. Further, I thank Peter Dizaj at United Custom Colors for his expertise.

No book is ever written entirely by the author: I have had the assistance of a wonderful group of editors: Josie Masterson-Glen and Steve Bliss from JLC have made this a personal and pleasurable experience; Rick Mastelli, whose assistance in this project came late, has had the greatest influence and I'm thankful for having met him — on the phone and through the ether.

I thank Kevin Ireton for sharing his experience, advice, and guidance. Without Kevin's friendship I might not still be writing.

And last but mostly, I thank my brother Larry for his indefatigable help both in and out of carpentry, and for his inexhaustible patience.

Gary Katz

Contents

Introduction xi

Chapter 1	Finish Carpentry and Design	.1
Chapter 2	Materials	.13
Chapter 3	Tools of the Trade	.25
Chapter 4	Exterior Doors	.43
Chapter 5	Windows	.69
Chapter 6	Interior Doors	.79
Chapter 7	Casing	.99
Chapter 8	Baseboard	.115
Chapter 9	Closet Shelving	.125
Chapter 10	Crown Molding	.141
Chapter 11	Decorative Walls	.159
Chapter 12	Decorative Ceilings	.181
Chapter 13	Decorative Doorways	.197
Chapter 14	Bookshelves	.219
Chapter 15	Mantelpieces	.237

APPENDICES

Appendix 1	Suppliers, Products, and Resources	.255
Appendix 2	Takeoff Charts	.259
Appendix 3	Crown Molding Miter & Bevel Angle Setting Chart	.263

Index 267

Introduction

In my family, working with your hands wasn't considered a smart thing to do, so I didn't choose carpentry as a career. Rather, after graduating from college, I backed into carpentry without looking. Although I enjoyed the work, and tried to do it well, I didn't have much respect for the job, and not just because I was using a shovel in those formative years more than a saw. I was reluctant about being a carpenter because in this country craftspeople — like teachers — do not enjoy the respect they richly deserve. I found it hard to overcome that attitude and fought to hold my head up to the bankers, doctors, lawyers, and PhD's for whom I worked.

But during the last few years, all that has changed. While researching a labor of love—a book I've been writing on mantelpieces — I've spent hours in libraries and days in historic homes. I've held in my hands 18th-century books on carpentry and architecture; I've admired workmanship that stands unrivaled even 300 or 400 years after being touched by a tool. Both experiences have taught me to respect this craft and the millennial legacy of carpenters to which we all belong.

One experience especially stands out. Good fortune allowed me the opportunity to visit the Rare Book Department at the Huntington Library near Pasadena, California. I was sure all the scholars scattered around the room — their heads bowed over aged volumes — could hear my every step across the vinyl floor as I approached the librarian's desk. When he kindly leaned out over the wide counter to pass me the first edition of William Kent's *The Designs of Inigo Jones*, two quite heavy tomes, I froze. He grimaced, frowned at me, and (the small of his back having had enough) dropped the books on the desk with a bang. Red-faced, I picked up the seminal pattern books on Palladian architecture and quickly found an isolated corner. When I opened the first volume, the list of "subscribers" jumped right out at me.

In the early days of publishing, books were often presold to subscribers to help defray the cost of publication. The subscribers names were frequently published in the book — just like having your name printed on the list of contributors to a charity. But in this list, the subscriber's professions were also included, one after another, and most were carpenters, joiners, carvers, and masons. Seeing that list changed my life. Those names of faceless craftspeople leapt out at me and reminded me of the woodwork I'd admired in countless historic homes across America. That list illuminated the admirable career I had chosen and the distinguished company of carpenters I was honored to join.

I've written this book for others who wish to be a part of that tradition, and I have organized the book after the pattern books that guided many 18th- and 19th-century carpenters. Pattern books always began with a review of the classical orders, and for good reason: those paradigms of Western style are the foundation for nearly every decorative element that a carpenter creates. In the first chapter, I'll explain more about the classical orders. In succeeding chapters I'll concentrate on efficient production techniques, but I'll continue to point out the influence of classical design. I'm convinced that, like me, most carpenters care about their craft and want to learn *why* as much as *how*.

Just the other day, while standing 30 feet in the air on a scissor lift at the peak of a 12/12 gable ridge, one of the fellows on our crew turned to me as we were applying a Victorian pendant and crown molding, and he said, "Learning something new almost every day is the sweet spot of this job." I hope this book will be a "sweet spot" for both new and experienced carpenters — for those who appreciate our nearly ageless craft. Countless times while working on this book, when the size of the task seemed overwhelming, a friend's advice came to mind: "Just take it one room at a time." I offer the same advice to fellow carpenters, whether they're on the job or visiting an historic home. I hope that readers will approach this book the same way I did — one room at a time — and that your enjoyment of carpentry will be as enhanced by the experience as has mine.

CHAPTER 1

Finish Carpentry and Design

The first assignment for most apprentice carpenters—after learning to sweep sawdust and carry material—is installing baseboard molding. An experienced hand usually offers a brief introduction to the basics of using a tape measure, and then a longer lesson on the use and danger of a power miter saw. Graphic and gory details often accompany this intentionally terrifying lecture.

Afterward, the shaken apprentice is led to a back-room closet and told with all the blunt force of a hammer blow: "Base that. When you think you're finished, come and get me."

I don't imagine that anyone has ever told an apprentice that they're about to install a molding that originated on Greek temples. In fact, most carpenters never learn that baseboard molding has its origin in the plinth and base of a classical column, or that modern casing and crown come from the same source. I can admit that many years went by before I learned that carpentry had a history.

My "education" began years ago, when I made two memorable mantelpieces for a client whom I'll never forget. Gladys surprised me from the start. The first time I visited her home there was a tall pile of stone, brick, and rubble in the driveway from the two fireplaces she had chipped clean, mostly by herself, though her husband looked pretty beat up, too!

Gladys wanted me only to "help" her design and build both mantels, so she said. With an inexpensive little power miter saw — and several boxes of finish nails in different sizes — she just didn't have all the tools she knew might be needed. But she sure had a pile of moldings, each one hand-picked because of its size, shape, complexity, and attractiveness. She even had a 12-ft. piece of rabbeted stair nosing. (Gladys had no idea where or how that piece would fit; she knew only that she liked it.) We ended up using the stair nosing and many other pieces, too. I would later learn that she had no hesitation about visiting that molding supplier again and again, even though it was a twenty-mile drive through traffic.

That job turned out to be real fun — like first love. Nothing we could do, no combination of moldings or styles, seemed wrong. I'd hold two pieces together and she'd say, "Yes, but how about this?" Then she'd hold those two pieces alongside two more and I'd say, "Okay with me."

In the end, those mantels knocked me out. I might even have a picture of them somewhere, but I'd never show it today. They were both silly flights of fancy. We had fun combining different moldings, but our lack of familiarity with historic design limited our effort to a tasteless conglomeration of coves, fillets, wave moldings, egg-and-dart ogees, more fillets, coin molding, bead-and-barrel patterns, and dentil blocks.

After building those mantels, I knew something was missing, and I was a man on a mission. I borrowed books on mantels from the library and looked at every magazine I could find. I learned that carpentry actually had an history, though no known date of origin, and that the moldings and designs I built were inspired by historical precedents, often by the "classical orders," whatever they were. While I searched for clues to help me understand mantelpiece design, I kept running into that "classical orders" phrase, especially in captions beneath photographs. But I had to look much harder to learn the real meaning of the words.

The job of carpentry hinges on more than just

woodwork and joinery. Building a successful mantelpiece, decorating a doorway, choosing wainscoting and crown molding, all depend on a carpenter's familiarity with his craft — which also means a familiarity with classical and period designs. There's no better place to begin that education than by learning about the classical orders; after all, they're behind almost everything a carpenter builds.

You might say that carpenters bring order to the world, by building places and spaces to better organize, protect, and provide comfort for people's lives. But carpenters also are dependent on order for the structures that they build. It's probably easiest to understand what classical orders are by taking the words one at a time. In architecture, the word *order* really means any horizontal structure that's supported by vertical structures. The relationship between an entablature and its supporting colonnade, for instance, represents an order. Even a simple doorway is an order, so too are a pair of posts supporting the header of a patio cover. The term *classical* refers to the art and architecture of the Greeks and Romans. So the *classical orders* are simply the three Greek orders (Doric, Ionic, and Corinthian) and the two Roman orders (Tuscan and Composite) that regulate the relationship between vertical and horizontal components in the architectural structures that are among the roots of our culture.

As I learned on those two mantelpieces years ago, it really helps to know whether an overmantel and a cartouche belong on a Federal or Georgian mantelpiece; whether Gothic and Celtic ornaments belong on a Victorian or Colonial mantel; or if it's acceptable to decorate an Arts and Crafts mantel with garlands and urns. It also helps to know if the dart on egg-and-dart molding should point up or down.

A PRIMER ON HISTORICAL DESIGN

This book isn't about classical design. It's about finish carpentry. So this short introduction won't last very long. I only want to point out a few of the valuable lessons I've learned about the way classical designs affect our craft.

In this chapter I'll demonstrate several ways that classical designs have influenced period styles. I'll also offer a few tips to help distinguish major architectural periods — using the vocabulary of a carpenter. In the following chapters I'll describe many of the modern materials, tools, and techniques currently used to achieve those designs.

For a better understanding and a richer appreciation of finish carpentry, I encourage readers to visit libraries and haunt historic homes. I can't imagine a more perfect vacation for a finish carpenter! After all, learning *why* is as important as learning *how*. Among the books I've found especially helpful are Cyril Harris's *Illustrated Dictionary of Historic Architecture* (Dover, 1977); *The Elements of Style*, edited by Stephen Calloway and Elizabeth Cromley (Simon and Shuster, 1991); and *A Field Guide to America's Historic Neighborhoods and Museum Houses*, by Virginia and Lee McAlester (Knopf, 1998). See also their book *A Field Guide to American Houses* (Knopf, 2000). For more intrepid readers/carpenters, I also recommend two histories of architecture: *A History of Architecture*, by Spiro Kostof (Oxford, 1985), and another that is out of print but can still be found at used bookstores and online, also titled *A History of Architecture*, by Fiske Kimball (Harper Bros., 1918).

Pattern Books

As carpenters collect tools and build on their skills, they should also collect and build a simple library of their craft — books, magazines, photographs, and drawings of historic and modern styles, effective proportions, and pleasing joinery. If you're like me, you'll begin collecting on subjects that interest you and later on add subjects that will interest your clients.

During the eighteenth and nineteenth centuries, pattern books were the only academic education for countless craftsmen and architects, reinforced by a long apprenticeship in the trade. Pattern books were similar in some ways to the piles of magazines and books published today which are devoted to "building plans" or "home decoration." But pattern books were also different. Mostly slim volumes, pattern books frequently began with elaborate illustrations of the five classical orders, and then continued with precise drawings and measurements which detailed designs contemporary to their publication. Stairs, doorways, windows, arches, even whole floor plans, were often included, and each in the style of the book. For instance, Asher Benjamin's *The Country Builder's Assistant* (1797) includes dimensions and details for each of the classical orders, but also plans and specifications for Federal-style mantelpieces, stairways, doorways, roof plans, and floor plans. Pattern books were devoted to other specific styles, too. Batty and Thomas Langley's *Gothic Architecture* (1742) begins

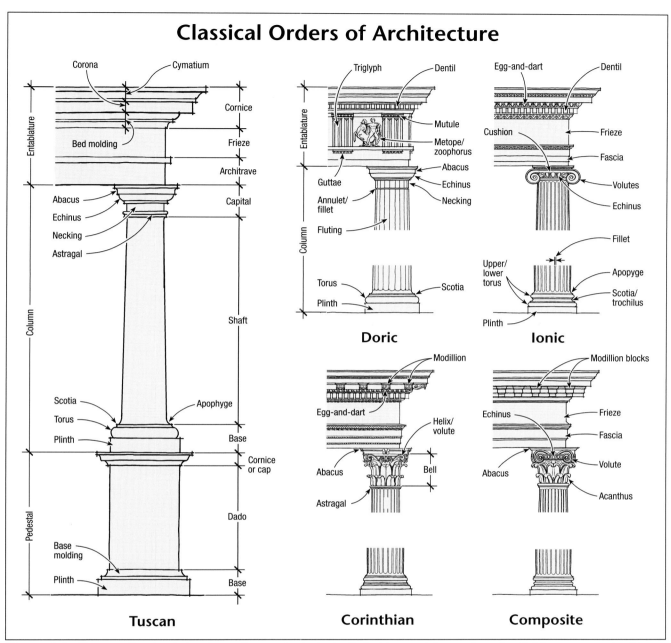

Figure 1-1. *There are five classical orders — Tuscan, Doric, Ionic, Corinthian, and Composite — but finish carpenters should try to recognize at least the Doric, Ionic, and Corinthian. These three form the basis of most designs you'll see today. The Doric shares several features and proportions with the Tuscan; the Corinthian shares features with the Composite.*

with elaborate and unusual details for five "new Gothick Orders," and continues with doors, windows, and facades all in the Gothic tradition.

The Five Orders

The classical orders tell a tale of history themselves because they're a combination of primitive, Egyptian, Greek, and Roman influences (*Figure 1-1*). For thousands of years, we've been stacking rocks and trees to build shelters, which partly explains why classical columns are shaped like tree trunks. Many capitals even have branch-like stems, some with leaves.

Each classical order is defined by strict proportional rules: The height of the column, including the base and capital, is determined by the width of the column near the base, as is the height of the entablature and the proportions of its components: the architrave, frieze, and cornice. For instance, Abraham Swan (following the Renaissance architect Palladio) wrote that the height of the Doric column

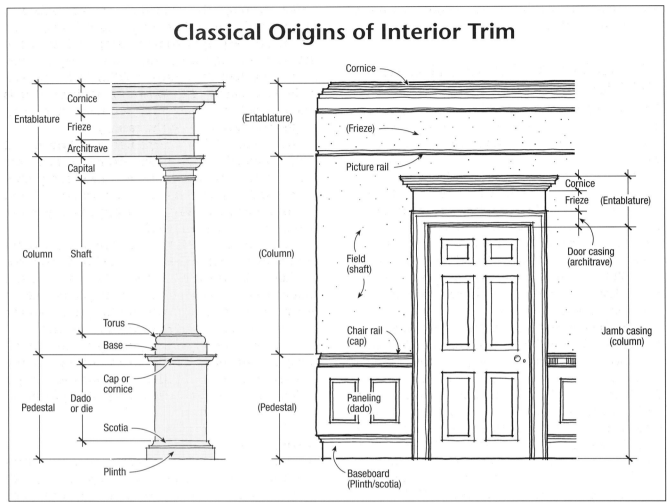

Figure 1-2. *The classical orders are the origin of baseboard, casing, wainscoting, crown molding, and other details, such as the components of a mantelpiece. In fact, classical terminology is still used to identify most trim elements.*

should be ten times the diameter of the column at its base, the height of the Ionic column eleven times the diameter, and the height of the Corinthian column twelve times the diameter.

The entablatures in all three orders, on the other hand, are the same size in relation to the diameter of the columns. Needless to say, when you vary the proportions of major structural components that much from one style to another, the effect on the details is considerable: There's an increased vertical *and* horizontal difference from the Doric through the Ionic to the Corinthian orders — toward more lift, poise, and decorative freedom. Given these dramatic differences, it's not surprising that such a great variety of contemporary styles have evolved from the classical orders.

Seeing Walls as Columns

As carpenters, we continue to build with the bones of classical architecture, though Palladio's strict rules are only a guide for modern design.

In 1897 Edith Wharton wrote: "It may surprise those whose attention has not been turned to such matters to be told that in all but the most cheaply constructed houses the interior walls are invariably treated as an order." Wharton, in her collaborative work *The Decoration of Houses*, goes on to describe how baseboard represents the plinth on a column, how crown molding resembles a column's cornice, and how the intervening wall — often decorated with paneling — mimics the shaft of a column *(Figure 1-2)*. Doors and windows, with their casing and occasional overdoor treatments, can be viewed as colonnades. These classical forms are easily identified in nearly every major period of American architecture.

AMERICAN ARCHITECTURAL INTERIORS

Recognizing classical elements in American architecture helps to identify one period from another, a bit of knowledge I've found helpful as a finish carpenter.

Not long ago, I was asked to install a mantelpiece for some clients who were building a Colonial home. The couple was interested in authentically imitating the period, but they were also attracted to reclaimed materials. They had salvaged the doors, windows, and hardware from an early 20th-century home, and it all fit fine with their desire for tall plinth blocks, rosettes, and overdoor crown moldings.

But the medieval mantelpiece they were thinking of purchasing had chamfered pilasters decorated with a fleur-de-lis design, and the frieze was emblazoned with a heraldic shield. I recognized the Elizabethan-Gothic style and knew it to be incompatible with the Colonial look they were creating. A small collection of mantelpiece details copied from historic homes convinced them, and also provided me with a guide to designing their mantelpiece.

Historic homes often have been a guide for me; they're the temples of our craft. Well-wrought homes display a harmony of design that's not easily forgotten. I'll never forget my first visit to the John Brown Home in Providence, R.I. When I walked through the front door and came face to face with the imposing entry and hallway, I lost my breath. My eyes couldn't stop taking in the details, from big things (such as the way the carpenters had cut the crown molding in the broken pediments above the doorways) to little details (such as how the chair rail terminated and overlapped the back edge of the casing).

But when fellow carpenter and friend Jed Dixon pointed out the baseboard profile and how it coped perfectly into the same matching profile on the radius plinth around each column, I was stunned. I couldn't wait for the opportunity to use that same technique myself.

The photographs in the following sections come from both historic homes that I've visited and modern homes in which I've worked. The combination should help to illustrate how a carpenter's life is dependent on both the past and the present — the first for motivation and the second for means.

This small gallery of images depicts only broad categories of architectural styles. Though many periods overlap and influences abound, major periods are marked by key distinguishing features.

Georgian

The Georgian period was named after the 18th- and 19th-century Hanoverian kings of England, including George III (the English king Americans rebelled against in 1776). Carpentry in the Georgian style can be identified by several noticeable features: tall doorways decorated with open

Figure 1-3. *The skirts on Georgian stairways (those small triangular spaces are known as "spandrels") were frequently hand-carved with S-curve decorations. Because casing is also known by its classical name — architrave molding — the design over the far doorway (and the one in the next photograph with ears in the upper corners) is called a "crosette architrave."*

Figure 1-4. *The carpenters used the same crosette architrave detail on the doorway to this parlor as in the previous photo. The mantelpiece has a full overmantel, with pilasters and a cornice. Both designs identify the style as Georgian. Notice the wings or ears at the ends of the mantelshelf, and how the cornice molding above the overmantel wraps ("breaks forward") around the upper pilasters. Those ears and breaks are also typical of the period.*

Figure 1-5. *Several features in this room help identify the woodwork as Georgian-style: Ionic columns support the spectacular mantelshelf and overmantel, while pairs of Corinthian columns carry the upper entablature, which is crowned by an open pediment. But notice, too, the heavy Greek-key dentil molding in the room's cornice, and the large brackets that act as keystones in the peak of each arch. All that luxurious detail is characteristic of the Georgian period.*

pediments, rich Ionic and Corinthian columns, and elaborate cornice work. This cornice work includes tall entablatures with bold frieze designs, large, deep dentil blocks, and sometimes modillion blocks, too (like a row of little corbels suspending the cornice).

Just as early American colonists relied on England for higher education, Georgian woodwork depended on tradesmen trained in English and European methods. Along with strict classical proportions, there's even a touch of the Rococo in many Colonial homes, such as the S-scrolls in the stairway skirt *(Figure 1-3)*.

Georgian mantelpieces are known for having large dramatic overmantels (almost a second mantelpiece or heavy framework above the actual fireplace). And crosette architraves — casing or a picture frame with ears in the corners — are common around doorways and above mantelpieces in Georgian homes (*Figures 1-4* and *1-5*).

If historic homes are temples to the craft of carpentry, then Georgian homes are our cathedrals. Wood carving was at its peak during the Georgian period (the work of one famous carver, Grinling Gibbon, was mimicked by many others), and carpentry was king. It's no wonder that other carpenters and I get dizzy visiting one of these homes; there's so much to see — both high and low — and never enough time to catch it all *(Figure 1-6)*.

Federal and Neoclassical

During the late 18th century and the early 19th century, people tired of imposing pediments above doorways and monumental overmantels.

Archeological digs during the mid 18th century had unearthed fabulous Roman ruins at Pompeii and Herculaneum. A "new" classical style was born, which was less imposing, grand, and monumental than the architecture of Georgian mansions, and was perfect for smaller, more intimate middleclass homes. Roman designs swiftly caught the imagination of English architects, especially the Adam brothers, and Neoclassicism emerged. In America the Adam style became a public fad, expressing all the romantic notions of a new republic *(Figure 1-7)*.

Large Georgian doorways and mantelpieces were rejected as carpenters collected new pattern books that arrived from England. Urns, garlands, vases, and arabesques of acanthus and honeysuckle stems were applied to simpler doorways, smaller mantels, and more delicate casings and cornice

Courtesy of John Brown House

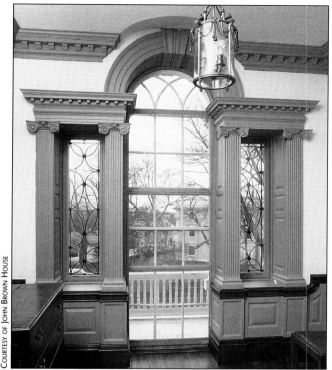

Figure 1-6. *An arched window flanked by two rectangular windows is known as a "Palladian" window — named after the Renaissance architect who inspired the Georgian style. Above the Ionic columns, a large dentil molding decorates the window's cornice, and little corbels support the room's cornice (those are called "modillion blocks").*

Courtesy of Trousdale

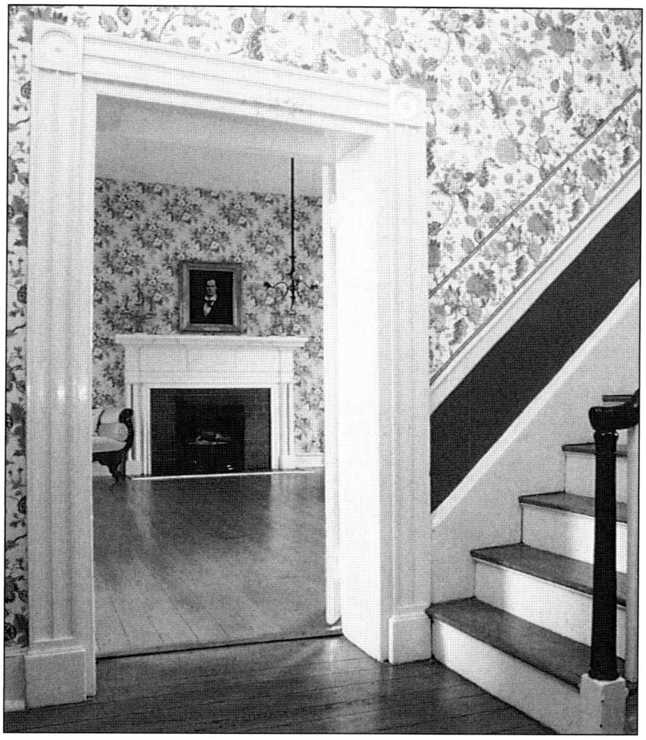

Figure 1-7. *Molding styles were more moderate during the Federal period, but carpenters were still active. The fluted casing, plinth, and rosette blocks identify the Federal period (also called "Neoclassical"), characterized by a renewed interest in classical architecture. This mantelpiece is known as a "simple" mantel because it has no overmantel.*

work ***(Figure 1-8)***. Rosettes and plinth blocks around doorways replaced full-sized pilasters and columns; heavy cornice work was rejected in favor of more delicate details borrowed from the Ionic and Corinthian orders. Modillion blocks were abandoned entirely. Details from the Ionic and Corinthian orders continued in use, but book-sized dentil blocks and fist-sized egg-and-dart moldings were reduced to postage-stamp and thumb-nail sizes that are still popular today.

Victorian

Like stock-market booms, fads always fall. The Neoclassical fever of the early 1800s burned out during the later part of the century and finish carpenters stopped depending on formal rules of classical proportion and style.

A grab bag of architectural styles emerged during the second half of the 19th century. These included a revival of Gothic designs, which lead to the Second Empire, the Stick, and the Shingle styles, and even later to the Aesthetic and so-called

Courtesy of the Gardner-Pingree House

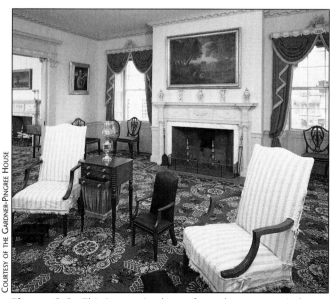

Figure 1-8. *This impressive home from the same period illustrates another aspect of the popular Federal style — the difference between a country home (previous photograph) and the home of an affluent ship owner. Still, the mantelpiece is "simple," and, the woodwork above the long doorway separating the two parlors doesn't include a pediment.*

Figure 1-9. *Non-carpenters often criticize the Victorian era as excessive. The spandrels in this stair are decorated with S-scrolls and the balusters are carved with pineapples. The narrowness of the panels increases their number and emphasizes the wainscoting. A wide and intricately carved plaster frieze complements the plaster crown molding. And the woodwork above the doorway — a sunburst-spandrel of spindle-work — may be elaborate, even effusive, but it's definitely not excessive for a finish carpenter.*

Figure 1-10. *Gothic designs were common in many Victorian homes. The pineapple pendants in the coffered ceiling, the arched niches flanking the dining-room doorway, and the cornice carving above the bookshelf (a lozenge pattern and a wave pattern) all point to the Victorian interest in Gothic designs. Meanwhile, the fluted casings and classical end blocks that flank the bookshelf openings — and the fluted casing, plinth blocks, and rosettes that surround the doorway — prove that the these carpenters had mastered the art of mixing period styles.*

"Queen Anne" styles. Carpenters no longer needed classical training or apprenticeship. Homes were decorated with a hodgepodge of twisted columns, spindles, balustrades, brackets, corbels, steep Gothic arches, heavy beams, fretwork, and lots of spandrels — miniature balustrades hanging from porch roofs and spindle-work sunbursts above doorways *(Figure 1-9)*.

But that wasn't all. The Victorian era was a carpenter's dream of woodwork featuring old-growth lumber, from redwood to oak, walnut to fir. Dark wainscot paneling, coffered ceilings with turned pendants, elaborate and imaginative overdoor pediments, mantelpieces with multiple stages and nooks and shelves for collectibles: all were common in many late 19th-century homes *(Figure 1-10)*.

Though the style has never been my favorite, like any finish carpenter, I'm always excited by the prospect of working on a Victorian home, not only because of the job security (there's always an endless amount of moldings to install), but because of the dramatic end results.

Arts and Crafts

Before long, the Victorian fad fell from favor, too, though carpenters suffered no loss. During the late 1800s, the industrial revolution picked up steam, first in Britain, and the simpler agrarian way of life was under pressure. But just when value seemed measurable only in terms of money, a return to handcraft erupted in England. The Arts and Crafts movement was a vigorous and passionate protest against thoughtless "progress" and quickly spread to America. Victorian gingerbread, clutter, and excess ended.

The Arts and Crafts period proved a heyday for carpenters. Architects adopted simple designs,

Figure 1-11. *Most carpenters favor the Arts and Crafts period and for good reason: The style celebrates hand-crafted workmanship and fine joinery. Some defining elements include Japanese motifs, like the steps cut into the beam ends (called a "cloud-lift" pattern), the cloud-lift corbels, and the same pattern in the rails behind the bench.*

Figure 1-12. *Oversized finger-joints, highlighted by eased edges, draw attention to the soft curving design and the artful joinery. Small pegs, sanded slightly proud of the surface, are a signature of the period.*

Figure 1-13. *Arts and Crafts architects favored cozy Gothic-style fireplaces, too, because they symbolized the importance of family, hearth, and home. This "inglenook" with seats on both sides is framed by a Gothic arch.*

mixing Medieval, Asian, and even a few Victorian elements. Throughout the period, fine woodwork — often exposed joinery — highlighted every design *(Figures 1-11 and 1-12)*.

Victorian mantelpieces, with their ornate ornamentation, were replaced by hand-wrought woodwork, often with Gothic references and mixed with other natural materials including iron, tile, stone, and concrete *(Figure 1-13)*.

Victorian coffered ceilings continued in use, in some cases with pendants and deep crown moldings, but doorways and casings reflected structural form more than decorative fancy. Post-and-beam construction — emphasizing mortise-and-tenon joints — appeared, while casing and baseboard were reduced to their simplest form, often applied with exposed plugs, bungs, or dowels accented by a different wood type.

Like every other architectural "reaction," many aspects of the previous period remained in use, which is one reason why parlor mantelpieces and narrow overmantels with mirrors were still popular on many Arts and Crafts homes *(Figure 1-14)*.

Modern

It has taken me many years to develop an understanding—and even more years to develop a liking — for modern architecture. As a carpenter, modern trim-less homes used to seem cold, austere, and un-homey to me. But that's no longer the case. I attended a lecture at the new Getty Museum shortly after it opened. Richard Meier, the architect of the museum, was the speaker and he began the lecture by answering the oft-heard complaint that the museum stuck out from the site like a sore thumb. Meier explained — as if he were speaking only to me — that on the contrary, he had designed the complex to fit into the land, to emerge from the Santa Monica mountaintops like a structure from classical antiquity, not only as a part of nature but also as an expression of our past and present culture.

At that moment I finally recognized the rela-

Figure 1-14. *Nearly simultaneous with the Arts and Crafts Movement, Beaux Arts homes were also a mixture of Gothic and Classical designs. The huge brackets supporting the mantel — carved with classical acanthus leaves — could be identified as Georgian if it weren't for the keystone-shaped medallion and Celtic shield, both medieval ornaments. Georgian carpenters would never have mixed these designs, but after the Victorian age, anything became possible.*

Figure 1-15. *Even Southwestern designs are dependent on woodwork. Doors and mantelshelves in fir, oak, or pine are common. Occasionally baseboard is treated similarly, but seldom casing.*

Figure 1-16. *Some critics think that new, modern styles — with white walls, steel posts and beams, and sharp angles — are cold and stark. In this home, however, austerity highlights fine woodwork, whether in a floor, ceiling, mantelpiece, or cabinetry.*

tionship between modern and classical architecture: Modern designs are an architectural reaction to the Arts and Crafts movement. They're a stripped-down version of classical architecture where posts and beams (columns and entablatures) are laid bare and nakedly support our protective shelters *(Figure 1-15)*.

Modern styles began during the Arts and Crafts period, but the movement really took off during the early 20th century when Walter Gropius and Mies van der Rohe started the Bauhaus school in Germany. The German society needed to rebuild after the devastation of World War I, and all decorative elements — eaves, cornice work, moldings, and even gable roofs — were discarded in favor of straight clean lines, flat roofs, and simple Greek-temple-like structures *(Figure 1-16)*.

These tenets of simplicity, mixed with the ageless influence of classical architecture (and the continuing desire to open our homes and our lives to the outdoors) are the foundation for many of today's modern styles. From straight baseboard with no molding profile to jambs without casing and

Figure 1-17. *Just as in the Victorian period, rules or strict distinctions between styles are often overlooked in today's homes. Contemporary carpenters face combinations such as light maple woodwork (coffered ceiling, crown moldings, and cabinets) mixed in the same room with a medieval mantelpiece.*

Figure 1-18. *The simple Craftsman Style continues to enjoy a deserved popularity, especially among woodworkers. Three types of wood were used to decorate this room in the home of a finish carpenter: The mantel is hickory, the floor and window casing are cypress, and the entryway (not shown) is Honduran mahogany.*

doorways without headers, from invisible European hinges to floating stair treads and walls of frameless windows, modern homes are a challenge, not a threat, to finish carpenters (**Figures 1-17** and **1-18**).

CARPENTRY TODAY

Modern finish carpenters install a broad variety of woodwork. The limited number of examples in this section point to only a few variations. I've worked in homes that have adhered strictly to a single historic style, but more often I've worked on projects that combine elements from several styles (sometimes tastefully, sometimes not).

No doubt, carpenters today seldom enjoy the respect once granted tradespeople. And there are many over-the-top jobs where a carpenter shouldn't dream of offering design advice — not when architects, designers, and decorators have already been enlisted for just that purpose.

But for average everyday projects, the role of a carpenter hasn't changed for hundreds of years: Carpenters are the first, last, and often the only word when it comes to woodwork design.

Which explains why a carpenter in search of success — one who cares about pleasing clients and making a profit — should learn the basics of period styles. After all, there are few completely new things under the sun, and our carpentry creations are all copied, borrowed, or suffused with something that predates our efforts and often ourselves.

Chapter 2

Materials

Though finish carpenters continue to use some of the same materials as our forbears, most of the materials we use today have changed dramatically, and materials continue to change almost daily. Change is an active principle in our trade, and maintaining a familiarity with contemporary materials has become the responsibility of every competent carpenter.

MOLDING

Foot for foot, the majority of the material finish carpenters install falls under the category of molding, including baseboard, casing, crown, paneling, and chair rail. Suppliers stock all of these moldings in a variety of different materials.

Stain-Grade Wood

Fortunately, natural wood hasn't lost its appeal, and carpenters are still blessed with the opportunity to work with solid wood species that have been popular for generations, including pine, fir (both face- or flat-grain and vertical-grain), mahogany, oak (flatsawn, quatersawn, and rift), cherry, walnut, and maple. But supplies of tight-grain old-growth lumber have diminished, and prices have risen rapidly. Like it or not, technology has again come to our "rescue" with the generation of several new "species" of trim, and today many homes are decorated with modern paint-grade products, some based on wood and some not.

Medium-Density Fiberboard

Medium-density fiberboard (MDF) is now as common as the flu and nearly as dreaded. Made from a stew of wood fibers, resin, and wax that is compressed under intense pressure, MDF is easy to cut and shape with carbide tools, and it paints beautifully, which explains its popularity. But the sawdust is an irritant to lungs and eyes. And MDF is rock-hard, which means that you have to screw it or nail it with a pneumatic nailer, in which case the surface mushrooms around the nail, and each mushroom must be chiseled or sanded flat before you can fill the nail hole and paint the surface *(Figure 2-1)*.

In addition to these generally acknowledged concerns, an ugly rumor persists that MDF molding is more prone to movement — both swelling and shrinkage — in those parts of the country with seasonally high humidity. Some contractors on the East Coast have told me that MDF is not marketed in their areas because of movement problems. But manufacturers have informed me that MDF mold-

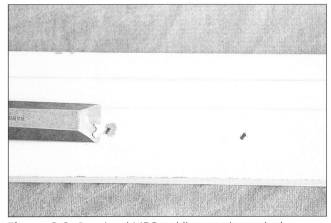

Figure 2-1. *Preprimed MDF moldings are increasingly popular because of their lower cost, ease of installation, and glass-smooth finish. But air-driven nails cause the surrounding surface to mushroom, and each mushroom must be chiseled off before filling the nail hole and painting.*

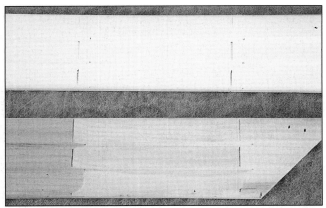

Figure 2-2. *Finger-jointed moldings have been marketed for more than a decade, and they're dependable if back-primed; without back-priming, the joints have a tendency to telegraph through the finish at a later date.*

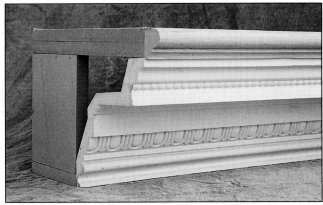

Figure 2-3. *Urethane molding is still fairly new but is rapidly gaining appeal, especially for large cornice work and architectural ornamentation (brackets, columns, medallions, etc.). A dramatic mantelpiece can be assembled from a single piece of urethane crown molding.*

ings are more stable than wood. That confidence may be justified partially because MDF moldings are always primed on both sides by the manufacturer.

My experience supports the manufacturers' statements. We install thousands of feet of MDF baseboard, casing, and crown every year. Even in homes that were completed eight and ten years ago, there is no evidence of material movement in our glued miter joints (and we miter all inside corners, both baseboard and crown as shown in Chapters 8 and 10).

Finger-Jointed Moldings

Before MDF moldings appeared in my area, finger-jointed materials were a popular and less expensive choice for most common moldings as well as jamb stock. Finger-jointed pine is significantly lighter than MDF, nail holes in it do not require special attention before filling and painting, and cutting it doesn't produce noxious sawdust. However, some of the individual finger-joints may require sanding. In addition, because each stick of molding includes a multitude of individual pieces, any swelling or movement can cause the joints to telegraph through the paint at a later date, so the material should be back-primed before application *(Figure 2-2)*.

Urethane Foam Moldings

One of the most popular manufacturers of polyurethane foam products is Fypon (see Appendix 1), and their large crown and cornice moldings have set the standard for the industry. Some of these styles cost as much as $50/ft., so they're not for every house or budget. However, a large dramatic overdoor cornice or a massive and intricate mantelpiece, can be achieved with a single piece of molding *(Figure 2-3)*.

Foam moldings are especially easy to cut and install. Most are applied with proprietary caulk or adhesive compounds. Unfortunately, the large crown moldings are available only in short lengths, from 7 ft. to 12 ft. (see Chapter 10, "Crown Molding").

Flexible Molding

Several types of flexible moldings (known as flex-molding or flex-trim) are available today *(Figure 2-4)*. Capable of conforming to any curve (I've wrapped flex-trim around 12-in.-diameter columns), flexible moldings have simplified the tasks of casing arched doors and running crown, base, and casing on radius walls. The least expensive flexible moldings are manufactured with a lower grade of resin, which doesn't flex as easily and can even be slightly brittle. Expensive grades of flexible molding are more pliable, have a smoother texture, and do not bruise or craze as easily from air-driven nails.

Stain-grade flexible molding with a grain pattern is also available. Though the material doesn't take stain any better than the smooth, paint-grade variety, the grain pattern does add authenticity to faux-finishes.

Carved Moldings

Though hand-carved moldings can be purchased or commissioned from individual craftsmen, the expense is often prohibitive. Machine-carved mold-

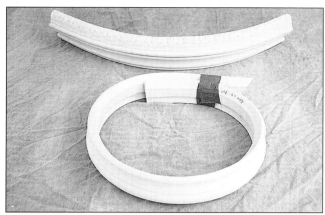

Figure 2-4. *When it comes to arched doors and radius walls, flexible moldings make a carpenter's job much easier and reduce job costs. Most standard profiles of casing, baseboard, crown molding, and even rabbeted panel molding, are available in flex-trim, but custom profiles can also be ordered.*

ing and ornaments, some produced in quantity, are a reasonable substitute. Raymond Enkeboll specializes in carved moldings that are manufactured by machine and finished by hand *(Figure 2-5)*. Though these materials are costly — some carved crown moldings cost more than $50 per foot — few manufacturers can compete with the availability or price of Enkeboll's products (see Appendix 1).

Embossed Molding

The greatest percentage of moldings stocked by suppliers are run through massive shapers, which cut with knifes. But some designs require more painstaking preparation, especially crown moldings that include egg-and-dart or bead-and-barrel decorations *(Figure 2-6)*. These styles, as well as many medallions and wooden ornaments, are both cut and stamped by presses.

"Compo" and Urethane Ornaments

Composition molding, a product that dates back to the 18th century, is still manufactured today in a similar mixture of whiting (chalk), resins, and glue *(Figure 2-7)*. However, "compo," as it is called, is fragile and somewhat difficult to install. Compo also has a relatively short shelf life and dries out quickly unless it is stored carefully. Dry and brittle compo is nearly impossible to apply.

In the last ten years, the use of composition molding has declined because of the emergence of urethane and resin ornaments, especially since molds are now made in highly flexible silicone

Figure 2-5. *Raymond Enkeboll has set the standard for carved moldings — from casing to columns — and continues to produce one-of-a-kind ornamentation.*

Figure 2-6. *Embossed moldings are stamped by machine and are far less expensive than carved molding. These mass-produced pieces — from baseboard to crown molding — add depth and drama to any project at an affordable price.*

MATERIALS 15

Figure 2-7. *"Compo" (composition molding) was derided in the mid-eighteenth century as a profane example of progress, but without compo many Federal-style homes would have been bare of ornamentation. J.P. Weaver, in Southern California, remains one of the few companies that continues to manufacture this clay-like molding.*

Figure 2-8. *Compo molding is extremely fragile and brittle when it dries, making it difficult to ship and store. Urethane moldings, more durable and of finer quality, have all but replaced compo. These examples are from J. P. Weaver's Petitsin collection.*

that can yield three dimensional ornamentation — leaves, fruit, rope — that outdo the flat figures allowed by older plaster molds. J.P. Weaver, still one of the nation's largest compo suppliers, now manufactures intricate, three-dimensional, carved moldings and ornaments in a resin material called "Petitsin" (see Appendix 1). Highly durable, extremely flexible, and easy to install, Petitsin moldings and ornaments do not dry out like composition moldings do. Petitsin designs enhance several styles, including Georgian, Rococo, Federal, Victorian, and contemporary mixed-period homes *(Figure 2-8)*.

MOLDING TYPES AND SIZES

Molding styles are divided according to categories, with the main categories being casing, baseboard, crown, and panel molding. Lumberyards carry a limited assortment of profiles in each of these categories, but well-stocked molding suppliers offer a large variety of off-the-shelf profiles in various sizes. Molding suppliers also stock unusual profiles (picture frame stock, stair parts, and such) as well as architectural ornaments (medallions, corbels, and such).

Casings

A carpenter's first chore in completing the interior of a new home is installing window and door casings, and molding catalogs begin with a selection of those materials. Used around window and doorjambs — both as an aesthetic trim to cover the gap between the jamb and the plaster, as well as an additional means of securing the jamb — casings range in size and design across the country.

On the West Coast, popular styles begin with inexpensive $1^1/2$-in. streamline and 711. This widely used profile is known on the East Coast as 356 or 351 *(Figure 2-9)*. Three-step, bead-and-barrel, and small fluted designs follow, also in $1^1/2$-in. sizes. These common styles are often available in $2^1/2$-in. widths, too, and the 60 percent increase in size is quite nearly matched by the increase in price. In fact, the size of the casing, not the design, almost always dictates the price. Universal casing in a $2^1/2$-in. size generally costs the same as other $2^1/2$-in. casings, such as 711, Colonial, Cape Cod, and 7-Bead.

On the East Coast, larger Colonial, Federal, Neoclassical, and Victorian styles are prevalent, along with a variety of rabbeted backboard moldings. These styles mimic the classical origin of window and door casings (see Figure 1-2, page 4). Many East Coast designs are less ornamental, yet the features are deeply incised, including acute quirks with proud beading.

The width of the casing in a home can often be limited by the size of the doorway returns. Efforts to reduce expensive square footage have diminished hallways, bedrooms, and bathrooms to their minimal sizes, and return walls flanking doorways are often framed only two studs wide. After drywall is installed, less than $2^1/2$ in. remains between most jambs and flanking walls, which limits the size of door casing. Fortunately, several $2^1/2$-in. and $3^1/2$-in. designs are also manufactured $1^5/8$- and 2-in. wide.

Full-service molding suppliers also stock an

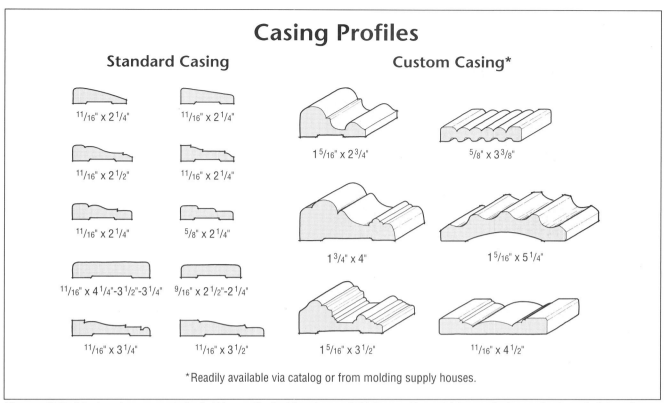

Figure 2-9. *Most casings are relatively thin on the inside edge and thicker on the outside, producing a shadowed picture-frame look around doors and windows. All casings should be eased on the outside edge to prevent splintering if someone brushes past.*

assortment of plinth blocks and rosettes, usually in a variety of sizes and designs. Fluted casings are frequently applied with plinth blocks and rosettes for a truly classical look, while Colonial plinth block and rosette designs depend on flatter simple stock combined with a casing that is decorated with a backboard.

All casing designs share several similarities. They vary in thickness from a thinner inside edge (the edge that is applied directly to the jamb) to a thicker outside edge, the effect of which is to produce the shadowed picture-frame look around doorways and windows. Also, casing is always rounded or eased on the outside edge, so that the edge won't tear or splinter if someone brushes or rubs past it.

Baseboard

Baseboard represents the plinth or base of a classical column (*Figure 2-10* and Figure 1-1, page 3). The height of the baseboard molding should provide a sense of structure and strength at the bottom of an interior wall, and that suggestion can be enhanced if the baseboard molding is taller than the width of the casing. But unless plinth blocks are installed, baseboard molding must terminate against the back of the casing, so the thickness of the baseboard molding should never exceed the thickness of the casing.

Unlike casing, most baseboard moldings incorporate a wide flat profile near the bottom (or thickest) edge, which allows space for carpet or other floor covering. The proud flat face of baseboard also protects the decorated upper profile from misdirected shoes, brooms, vacuum cleaners, and high-speed toys. While the outside edge on most casings is round, the bottom edge on baseboard molding is always cut sharp and square, because it abuts the floor.

Baseboard designs are nearly as diverse as casing, though it's the size of baseboard that varies the most, with stock designs ranging from 2 1/2 in. up to 8 in. Once again, smaller 2 1/2-in. streamline and 711 styles are prevalent in inexpensive homes. Taller baseboards, which incorporate a base cap detail within their profile, are more costly. Even larger 3/4-in. flat stock can be installed as baseboard, with the top edge finished by a separate base cap.

Such a base cap is especially handy for homes with wavy walls. Baseboard that is 3/4 in. thick is too stiff to follow a wall that isn't straight, but base

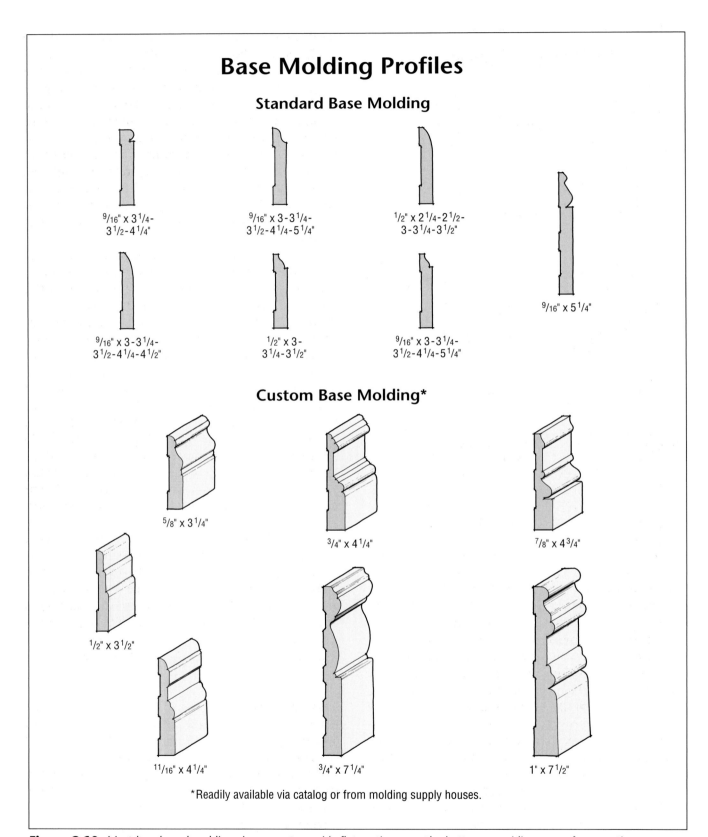

Figure 2-10. Most baseboard moldings incorporate a wide flat section near the bottom, providing space for carpeting or other floor coverings. A two-piece design using 3/4-flat stock with a separate base cap can help hide the irregularities of wavy walls in remodeling.

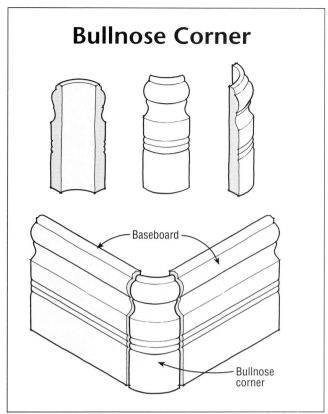

Figure 2-11. *Bullnose molding corners are manufactured in both stain- and paint-grade materials to simplify the trimming of rounded plaster and drywall corners.*

cap is more flexible and can cover gaps between the top of the baseboard and the wall, especially those too large to caulk.

Base shoe is installed at the bottom of the baseboard, particularly in rooms that are finished with hard-surface flooring such as wood, stone, or ceramics. Baseboard is too wide to follow an uneven hard-surface floor. Being small in section, base shoe is flexible and can follow a wavy surface, ensuring a tight, caulk-free joint with the floor.

Round or bullnose corners on interior walls, common in Mission-style homes with uneven, hand-plastered walls, are popular in many modern homes *(Figure 2-11),* and round drywall corner bead creates a perfect radius for every corner. Matching round corners are also available for many baseboard and base-shoe patterns, so that round corners can be finished quickly and perfectly. Bullnose corners are manufactured in both natural wood and paint-grade resin.

Crown Molding

Crown molding is installed at an angle across the corner where the wall meets the ceiling and imitates the cornice of a classical entablature *(Figure 2-12),* adding both period detail and architectural drama to the decor of a home. But choosing the proper size of crown molding can be confusing.

Historic rules of proportion determine both the design and the size of classical details. As we saw in Chapter 1, columns imitate trees, with the height of a column's shaft determined by its diameter. The height and width of the capital is also relative to the diameter of the shaft, and the height of the entablature is determined by the same dimension. The size of crown molding is therefore dependent on the height of the ceiling.

General practice for standard 8-ft. ceilings dictates that crown molding shouldn't cover more than 4 in. of the wall. Otherwise, the height of the room will be diminished and the walls will be dwarfed by the crown. Naturally, as the height of the walls increase, the height of the crown molding should also increase. Furthermore, for smaller rooms and lower ceilings it's also best to choose a crown molding that projects a little farther into the room, covering slightly more of the ceiling, so that the profile of the molding angles toward the eye. The profile of the crown molding should be visible without raising your head severely. In addition, most crown-molding details are cut at right angles, often with spaces between steps, so that the design can be distinguished even in rooms with low light.

The projection of crown molding is determined by its size but even more so by its angle. The angle of the crown, sometimes referred to as the spring angle, varies, but the two most common angles measure either 45 degrees or 38 degrees from the wall. Note: 45-degree crown molding is naturally 45 degrees from the ceiling, too, whereas 38-degree crown molding is 52 degrees from the ceiling (see Chapter 10, "Crown Molding").

Panel Molding

Panel moldings are used for a variety of purposes, and many molding patterns fall within this broad category of trim. Small edge bands used on common shelving and elaborate nosings used on mantelshelves are both basically panel moldings. But the category is best known for moldings that are applied directly to drywall (for simple flat panels) and rabbeted moldings that wrap around stiles and rails and create sophisticated recessed panels *(Figure 2-13).*

Several small pieces of trim also fall into the category of panel molding, including small ogee and cove moldings, as well as cloverleaf, beading,

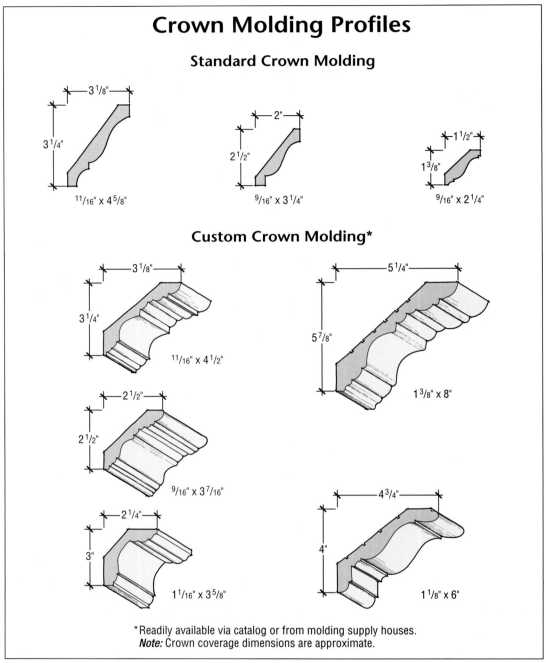

Figure 2-12. *Crown molding imitates a classical entablature and can add drama and period detail to a room. The molding size should be proportional to wall height — generally covering no more than 4 in. of wall with an 8-ft.-high ceiling.*

and astragal designs, all of which can be used by themselves or combined with other pieces to form elaborate molding schemes.

Chair rail is a specific type of panel molding that is usually applied at the top of a dado or wainscoting *(Figure 2-14)*. For that reason, some chair rails are rabbeted, so that they'll cover both the top edge of the paneling and the gap or joint between the paneling and the wall. Flat chair rails, some of which are reversible, are designed for walls without wainscot paneling and can be applied directly to the drywall.

SHEET GOODS

For several decades the term "plywood" described all sheet material—like the term Kleenex®, it was used for every similar new product that came on the market. That is, until particleboard emerged. Today there are many new types of sheet materials available, and "plywood" is used exclusively to describe sheet materials produced from multiple thin layers of wood. To ensure the stability of the panel the layers are cross-laminated: Each layer is coated with glue and oriented so that the grain in adjacent layers is perpendicular. Then the entire

Figure 2-13. *Many molding patterns fall within this broad category of trim and serve a variety of purposes. Rabbeted versions traditionally wrap around rails and stiles to create recessed panels, while standard profiles can be directly applied to drywall to create simple flat panels. Other uses include shelving edge bands and elaborate nosings on mantels.*

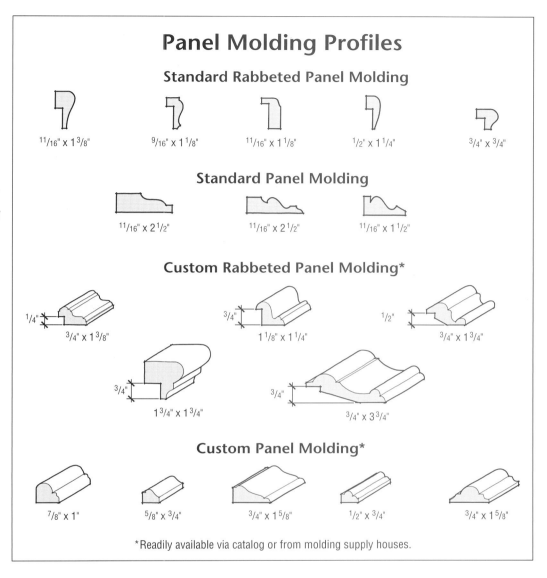

composition is compressed under high heat, which creates a strong and durable bond.

Maple "Apple-ply" and Baltic birch "multi-ply" are two popular plywood products used for fine finish work, especially cabinet drawers (at left in *Figure 2-15)*. Both types of plywood are made with fine, even-grained woods and with far more laminations than standard plywood; they can even be purchased with a lacquer finish.

MDF and MDF-Core Plywood

Not many years ago, the material of choice for paint-grade paneling, cabinets, and shelving was plain birch plywood (not the finely laminated Baltic birch plywood). An even-grained hardwood, birch can be painted to a smooth slick surface. Birch was also relatively inexpensive—at least up until several years ago.

Nowadays, MDF is the leading choice for paint-grade applications because it costs less than half the price of hardwood plywood and produces a glass-smooth painted or lacquered surface. MDF should not be confused with particleboard. Particleboard is a combination of sawmill shavings (coarse particles rather than fibers); it is formed under less pressure (and so is not as dense); and it cannot be painted to a smooth finish or machined easily the way MDF can. The edges of MDF shelving are easy to sand or rout with a beading bit, so there's no need for additional edge molding. Manufactured sheets of MDF are available in stock 4x8-ft. and 4x10-ft. sizes. Sheets that are 5 ft. wide and 12 ft. long can be special ordered, too.

Because of its stability and workability, MDF also has become popular as a core material for veneered sheets, and frequently replaces cross-laminated plywood. Nearly every type of veneer can be

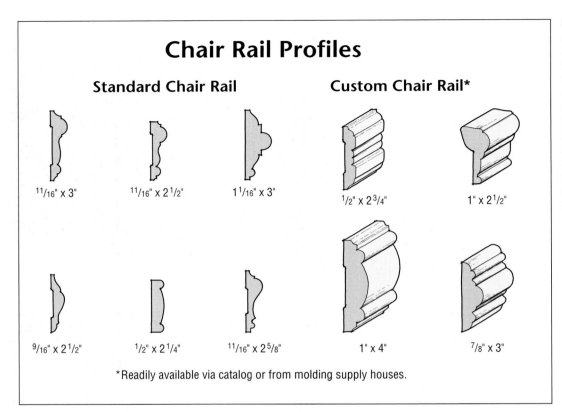

Figure 2-14. *Chair rail is a type of panel molding traditionally installed at the top of wainscoting. Some chair rails are rabbeted to receive the top edge of the wainscot paneling. Other profiles are flat and can be applied directly to drywall as a decorative element without any paneling.*

Figure 2-15. *Maple ply and Baltic birch (at left) are both made with numerous plywood layers, resulting in a strong, durable, and stable sheet. But these are relatively expensive compared with hardwood-veneered MDF-core (at right), which has become the most common sheet material used for stain-grade paneling and trim.*

purchased with an MDF core. This new species of sheet material is known as MDF-core plywood (at right in *Figure 2-15*).

Veneer Types

Veneer can be sliced from a log in several different ways. Each type of cut results in a different grain appearance.

Rotary cutting is the most common method for slicing veneer and produces the longest and widest sheets of material. The process is almost self-descriptive: A log is spun on a lathe while a long razor-sharp knife, set parallel to the axis of rotation, slices off a continuous sheet of veneer. Rotary-cut veneers have the wildest grain patterns and are used for general construction plywood but not for finish work. Rotary veneers form the core layers for most plywood products *(Figure 2-16)*.

Flat- or *plain-slicing* is a much slower process. A flitch or log is mounted on a table and sliced by moving the table repeatedly past a knife set parallel to the table. Each of the separate pieces of veneer is then applied, with edges abutting, to the face of a panel. The grain in each piece is similar to that of the preceding piece, and is tighter and

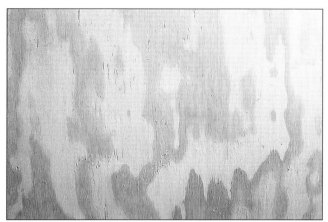

Figure 2-16. *Rotary-cut veneer is the most common type of veneer (pictured here in fir floor sheathing), and it's used mostly on construction grades of plywood because it doesn't finish well — with paint or stain.*

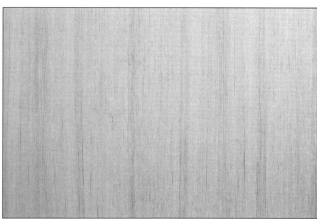

Figure 2-18. *Tight-grain stripes make it easy to identify quarter-sliced veneer (pictured here in Douglas fir and known as "vertical grain" or "VG" fir).*

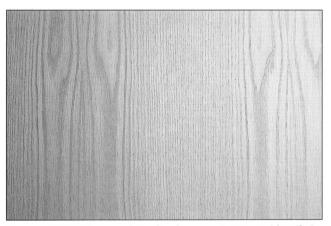

Figure 2-17. *Flat- or plain-sliced veneer is easy to identify in any species (pictured here in oak) because the sheet material often has the appearance of edge-laminated boards, like a plank-top table.*

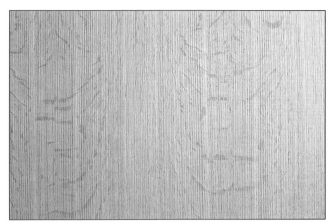

Figure 2-19. *Rift-cut veneer is available only in oak (pictured here in red oak) and resembles quarter-sliced veneer.*

more like that of sawn wood than that of rotary-sliced veneer *(Figure 2-17)*.

Quarter-slicing produces a straight, striped, and — depending on the wood species — closely spaced grain pattern, like that of vertical-grain Douglas fir. Quarter-sliced veneers are produced by mounting a quartered log to a table and cutting veneer at right angles to the growth rings. The grain pattern in the veneer is usually straight, and — depending on the age of the log — closely spaced. The unique rays in oak, which run like spokes from the center of the log, are best revealed by quarter-sawing *(Figure 2-18)*.

Rift-cut veneer is obtained exclusively from oak logs. To produce the most regular, parallel grain pattern in oak, rift veneer is cut at an angle of 15 percent off the quartered position *(Figure 2-19)*.

Flexible Plywood

Sometimes called "wally board" or "wiggle board," flexible plywood is available ³/₈ in. thick in 4x8-ft. sheets. It can be bent easily around radius walls and arches, and then used as an underlayment for a veneer.

Flexible plywood bends perpendicular to the grain direction of the face layers. To accommodate both wide walls and narrow long arches, flexible plywood is manufactured in two grain patterns, one with the grain running parallel to the length of the sheet, and another with the grain running perpendicular to the length of the sheet. Of course, each type still bends perpendicular to the grain of the face layers *(Figure 2-20)*. New types of flexible sheet material are rapidly appearing on the market. One product, called Timberflex®, is manufactured

Figure 2-20. *Flexible plywood is another welcome advantage for modern carpenters. Curved and radius projects — walls, arches, cabinets, stairs — are simplified by materials like wally-board, wiggle-board (pictured), and Timberflex®.*

with machine-kerfed particleboard backing and a smooth poplar face that can be painted, stained, or used as a substrate for laminating. Timberflex® bends smoothly, without dips and bellies, maintaining a fairly consistent radius because of the kerfed backing. Timberflex® is available in thicknesses up to 3/4 in. Even the 3/4-in. material can bend a radius as tight as 2 3/4 in.

Melamine and Kortron

While both these types of prefinished sheet goods often have particleboard at their core, there is a big difference in the surface finish. Kortron® is a white paint that is electronically baked onto the substrate. Melamine® is a low-pressure laminate: a vinyl-like paper is impregnated with melamine resins and then heat-pressed onto the substrate. Melamine® is a more durable finish and often is the choice for closet shelving and cabinet interiors (see Chapter 9, "Closet Shelving").

Choosing the right material for a job is often the first task for a finish carpenter, but finish carpentry is even more dependent on tools. In the next chapter, I'll describe the tools I use to cut and install all these materials.

Chapter 3

Tools of the Trade

This is a book on finish carpentry, not a tool review, so I won't attempt to wade through the multiplicity of tool models and manufacturers. Instead, I'll stick to what I know, the tools that I use and depend on every day.

But first, I have to repeat an overused phrase: Always buy the best tools you can afford. You already may have started collecting light-duty equipment, but before spending piles of cash on more "discount" tools, let alone replacements or repairs to them, remember that the amount of time devoted to learning a craft can be considerably reduced through the use of good tools. Good tools are not only a joy to use, but they also make for good work; poor tools guarantee poor results.

The brands and models of tools that I use and that appear in the following photographs are offered only as examples. Other name-brand manufacturers offer similar professional-quality tools. Nor is the following discussion of tools for finish carpenters meant to express any sense of priority. I began to buy tools as I needed them. Knowing what you need is the best guide toward buying better tools, though being prepared for the largest and most expensive investments never hurts.

MITER SAWS

Above all else, trim carpentry requires a good power miter saw, and there are a lot of brands to choose from. Which model is best depends on exactly what type of work you do. Put off purchasing this expensive item until you've developed a sense for the scope of your work.

Sliding compound miter saws are an attractive choice for finish carpentry because they can be used to cut wide boards: shelving, tall baseboard, and large crown molding, for example *(Figure 3-1)*.

Unfortunately, most sliding saws can't make a very deep vertical cut because of the small blade size. In fact, the highly popular Hitachi 8-in. sliding saw can't even cut 4-in. baseboard held vertically against the fence. Even 10-in. sliding saws can cut only the smallest of crown moldings "in position" — held against the saw fence at the same angle they will stand against the wall, though upside down. Instead, with a sliding saw you have to cut the material lying flat, swing the table to the proper miter, and rotate the motor and blade to the proper bevel. To cut the opposite miter and bevel, you have to swing the table to the opposite miter and flip the material end for end, or rotate the motor in the other direction (see Chapter 10, "Crown Molding"). Moving the material and the motor requires more time, more physical energy, and most importantly, more careful thought.

I use a Makita LS1211 sliding saw with a 12-in.

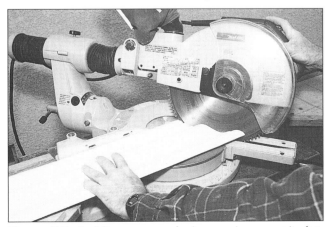

Figure 3-1. *A sliding compound miter saw is a necessity for cutting wide boards and large crown moldings.*

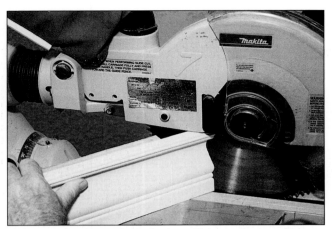

Figure 3-2. The saw's blade size will determine how big a vertical cut it can make. This 12-inch Makita sliding saw will cut 4¼-in. baseboard or crown molding standing up. Anything larger must be cut lying down.

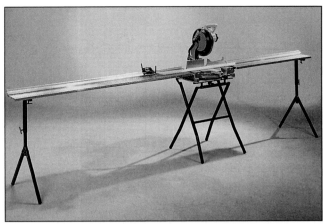

Figure 3-3. This portable miter saw stand from American Design and Engineering is rigid yet compact — you can carry it in your truck so it's always available.

blade, the largest blade size available on a sliding saw. I'm able to cut 4¼-in. baseboard standing up, with a homemade removable fence in place to support the material *(Figure 3-2)*. I can also cut most sizes of crown in position if those moldings do not exceed a 4¼-in. vertical coverage. For larger crown moldings, my saw tilts in both bevel directions. Of course, the saw is also ideal for cutting prefinished melamine and hardwood shelving.

Because it's a lot easier and faster to cut baseboard and crown molding when it's standing up than when it's lying flat, the saw of choice for our production crew is a 15-in. Hitachi C15FB. Though the Hitachi weighs considerably more than my Makita (80 lbs. vs. 50 lbs.), the Hitachi can cut baseboard and crown molding, in position, up to 6 in. tall.

Miter Saw Stands and Extension Wings

Even the most expensive sliding miter-saw is inefficient without a good stand. Operating a miter saw without a stand is like paddling a canoe with your hand instead of a paddle.

A useful and productive miter-saw system is dependent on several features: a good miter-saw stand; long, durable extension wings; and a repetitive stop system. I call this entire assembly a miter-saw work station (see "Making a Repetitive Stop System").

Several manufactured stands are available. Choose carefully. If you work primarily on a single job site for extended periods, and if you're able to secure your equipment at the job site, then a semi-portable stand with wheels might be the best choice. Many of the models, like the Rousseau, the Trojan, and the IDMM stand, allow for the saw to be permanently attached to the stand and then wheeled onto the job site. The extension wings and legs are then attached.

However, I'm rarely on a job site for more than a few days, and I can never leave my equipment, even if it's chained. My schedule can change from day to day, and on a given night I may not know what I'll need the next day. For that reason, I carry all of my tools in my truck, all of the time. Therefore, my miter-saw station has to be both portable and compact. American Design and Engineering (Appendix 1) manufacturers a rigid yet compact stand *(Figure 3-3)*. I had one of their early models for many years, until my truck was stolen. Because I had to buy so many other new tools, I couldn't afford to replace that stand. Instead I bought a simple, inexpensive, collapsible stand, and I borrowed several AD&E features when I built my own slightly lighter, wooden extensions (see "A Homemade Miter-Saw Station," page 28).

TABLE SAWS

When I was eleven or twelve years old, I sometimes went to work with my father and he'd leave me on the job site with the painter, the concrete crew, or the carpenters. Neal, my dad's finish carpenter, cast the spell of finish carpentry on me. Neal enchanted me with his craftsmanship. He built everything

Making a Repetitive Stop System

Finish carpentry involves repetitive work. A house might have 20 new interior doors, which means 80 legs of casing, all nearly the same length. Most of the doors will be the same width, too, while many of the windows will be the same size as well. Pulling a tape measure for each piece of molding takes too much time, and the risk of error increases each time you use a tape measure.

Many carpenters set up an easy repetitive stop by clamping a block to a long piece of 1x4 *(Figure A)*. While this setup works well, I prefer a system with more flexibility. I like having multiple stops for cutting legs and heads simultaneously. And I prefer a system with stops that are quick and easy to set up for even three or four cuts. Several companies manufacture good repetitive stop systems. Rousseau (Appendix 1) markets a lightweight, telescoping stop system (Model SS 6000) that attaches to their steel extension wings but can be mounted to any wing. It allows two separately adjustable stop positions. Because door and window sizes vary so much, however, I've found that more stops are better, so I made my own system. It consists of a fence that slides along a track at the back of the extension wings, and several sliding flip-stops *(Figure B)*.

The sliding fence. I made my fence by laminating two pieces of Baltic birch to a center strip of walnut. It looks neat and is lightweight, straight, and rigid. I cut a 3/8-in. dado down the center of each face, then cut a T-slot in the bottom of each dado using a table-mounted router fitted with a "T-slot" cutter (Appendix 1).

Next, I cut two pieces of 5/16 x 5/16-in. hardwood and attached them with small flathead screws along the top rear edge of my extension wings. This is the track, or runner, that my fence slides on. I could have made a long fence, but storing it in my truck would be a problem. I wanted to be able to cut molding for 8-ft. doors, so a sliding fence was the only alternative. T-bolts secure the flip-stops to the top of the fence, and a T-bolt mounted in each extension wing secures the fence to the wing.

The flip-stops. The repetitive flip-stops slide in the dado at the top of the fence. I rabbeted two sides of a piece of scrap hardwood, leaving a 5/16 x 5/16-in. tongue, then cut the long piece into shorter lengths to make several sliding base pieces. The 5/16-in. tongue rides in the dado, and a T-bolt, tightened from above by a turn knob, secures the bottom of the sliding base to the fence.

Figure A. *Extension-wing legs must be adjustable and portable. Steel is the best choice.*

Figure B. *A more flexible alternative consists of a fence that slides along a track at the back of the wings, and a pair of sliding flip-stops.*

I made the flip-stops from 1/2-in. plywood. I drilled 5/16-in. holes through each of the sliding base pieces, then used machine bolts and nuts with nylon inserts that would secure the flip-stops but still allow them to move.

A Homemade Miter-Saw Station

A miter-saw station consists of a table saw stand and a pair of extension wings with adjustable legs *(Figure A)*. I use an inexpensive folding table for a saw stand. I use one that was manufactured by Ryobi, although they don't make it anymore. Makita makes a good folding tool stand, and a Workmate will also suffice, but it will be a little on the heavy side.

The stand, however, isn't nearly as important as the extension wings. The wings not only help to support long pieces of material, but they also serve as an essential work station. They provide a waist-high, stable surface for pulling careful measurements. Extension wings also provide a durable solid surface for clamping material that must be cut with a jigsaw, or planed, sanded, drilled, etc.

Mounting the Wings

One important design element to consider when making extension wings is the mounting system *(Figure B)*. I wanted to take advantage of the manufacturer's wing-nut clamps, which secured the steel extensions that came with my Makita saw. Using a jigsaw fitted with a metal blade, I cut short pieces of dowels from the steel extensions, then mounted the dowels in blocks of clear two-by. I carefully located the holes in the blocks so that they lined up with the holes in the base of the saw, but I drilled the holes a little oversized, just to be sure of the alignment. I inserted the metal dowels into the holes in the saw base, then I filled the holes in the block half-way with epoxy and slid the block over the dowels. I let the whole thing dry for an hour. When I removed it, the alignment was perfect. All that remained was to screw the wings to the mounting blocks.

I made the wings 10 in. wide and 5 ft. long. The width provided plenty of working room and allowed for a repetitive stop system in the back (see "Making a Repetitive Stop System," previous page). The length was determined by the available space in my truck, the maximum length that would be comfortable and safe to carry, and the minimum length necessary for supporting long pieces of crown and baseboard. To keep the wings light and still durable, I used 1/2-in. Baltic Birch plywood. (I used the same material for the adjustable legs.) When I attached the wings to the mounting blocks, I found that each side was a little low of the saw table surface, so I inserted a plastic laminate shim between the wing and the block and raised each extension wing until it was perfectly flush with the saw base.

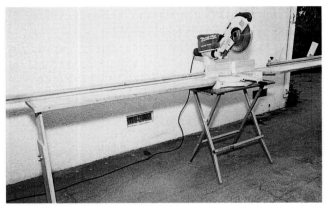

Figure A. *A good miter saw station can be made from a folding work table and a pair of site-made extension wings with adjustable legs.*

Figure B. *Short steel dowels align the extension wings to the saw.*

Another mounting alternative is to screw a large hinge to the base of the saw, then mount the mating leaf to the base of the extension wing *(Figure C)*. Only a removable pin needs to be pulled to break down the unit. This system is easy to install, and results in a stable, durable design. I prefer the steel dowel system because hinge pins can get lost.

To use the hinge-mounting system, always start by mounting the hinge to the wing first, or to a scrap of material equal to the thickness of the wing. Use this mock-up to precisely locate the hinge screw holes on the saw base. Install at least two screws through the hinge to the saw base. Drill and tap for those screws carefully. Through-bolts may also be used in some cases, if required. Minor adjustments are simple to make: Shims can be placed between the hinge and the bottom of the wing to raise it flush to the saw; or the hinge can

Figure C. *Hinges can also be used to mount the wings to the saw. Breaking down the unit is a matter of removing the hinge pin.*

Figure 3-4. *Most tool manufacturers make portable table saws that are ideal for finish carpentry. These range from lightweight $200 homeowner saws to heavy-duty professional models that cost over $550.*

Figure D. *Making the extension wings' legs adjustable will help keep them level on uneven floors.*

be mortised slightly into the base if the wing is proud of the saw base.

Adjustable Legs

I've also made adjustable legs by using two pieces of plywood for each leg *(Figure D)*. To keep the legs aligned as they are being raised and lowered, I cut a dado in each piece of plywood, 3/8 in. wide by 1/4 in. deep, then glued a 3/8 x 5/16-in. strip of hardwood in one dado on each set. I left the opposite dadoes empty as tracks. To secure the sliding legs, I mounted a 5/16-in. carriage bolt in one side, and cut a 3/8-in. slot in the mating piece. A large plastic knob (Woodworkers Supply, Appendix 1) cinches the two-piece assembly.

right on the job site. He made the cabinet cases, drawers, and doors; the jambs and door stops; the closet surrounds, doors, track, shelving, and supports. Neal cut nearly every stick of wood with a tremendous cast-iron table saw. That machine towered over me and even dwarfed Neal, who had to stoop to fit through standard door jambs. It took four men to move that saw.

Table saws are still a necessity for finish carpenters, but they're no longer four-man cast-iron leviathans. A variety of portable tables saws are available today *(Figure 3-4)* within three clearly identifiable categories: entry level, mid level, and top of the line. Entry-level saws are designed for occasional and part-time woodworkers. They cost $200 to $300 and weigh about 35 lbs. Mid-level tools are heavy duty and can take the abuse of full-time use, but they cost about $400 and weigh in at 40 lbs. Top-of-the-line saws are professional-duty tools, also. They start around $550 and often weigh a little too much to be carried easily by one person, especially if the distance from the street to the job is considerable. The advantage of a top-of-the-line saw is its ability to rip wider pieces of stock. Most have extending tables or fence systems that allow rips up to 25 in. — past the center on a sheet of plywood *(Figure 3-5)*.

However, I often need to rip and cross-cut 8-foot sheets of material plywood and medium-density fiberboard (MDF). For that reason, and because the entire package is easier to carry and requires less storage space in my truck, I use a mid-level Makita 2703 table saw in combination with a Rousseau Porta-Max table-saw stand. The collapsible yet rigid stand not

Figure 3-5. *Some professional-grade saws come with extending tables or rip fence systems. The one on this DeWalt can handle rips up to 25 in.*

Figure 3-6. *A good setup for cross-cutting 8-ft. sheet goods is a mid-level table saw, a collapsible table-saw stand, and a site-made outfeed table.*

only secures the saw at a comfortable height, it also increases rip capacity to 32 in.

The Makita saw cuts almost as smoothly as the top-of-the-line Bosch, Rigid, and DeWalt models, but it weighs 20 to 30 lbs. less and costs half as much. Adding a stand and accessory wings to the package makes the total cost slightly more than that of a top-of-the-line saw, but allows me to rip to 32 in.

Table-Saw Stands and Fences

Several saw manufacturers make stands with wide extension tables and large rip fences that improve the performance of even low-priced entry-level saws. Makita and Delta sell stands for their portable saws. Trojan manufactures stands for several saws. And Rousseau offers a universal stand that fits all entry-level tools and most mid-level saws, too, regardless of the manufacturer. (Rousseau actually manufactures the Makita stand.) I use a Rousseau stand with my Makita table saw, along with an outfeed table that I built from 1/2-in. Baltic birch plywood *(Figure 3-6)*. The table I built is 24 in. wide and 48 in. long. It attaches to the stand with a simple 1/2 x 1/2-in. hook-strip of wood, screwed to the underside *(Figure 3-7)*. Two legs are also hinged to the underside, beneath the back edge.

When working alone, an outfeed table prevents long boards and sheet goods from dropping off the back of the saw. Trying to hold a long board or a sheet of plywood down on the table after you've ripped more than half of it is difficult and dangerous. The use of a power tool should never require an unusual amount of effort. Whenever I'm about to push a power tool into that danger zone, I imag-

Figure 3-7. *A simple 1/2 x 1/2-in. hook-strip of wood secures the outfeed table to the saw.*

ine flashing lights and an approaching train. Like trains, power tools don't know you're there.

NAIL GUNS

When my father built homes in the 1950s, material was like gold and labor was cheap. Material is still expensive, but labor is almost priceless. Given that fact and the increasing use of MDF trim materials and urethane moldings—combined with today's tighter schedules—faster fastening methods are a necessity.

Many of the current nailers on the market are worthy tools, but before investing in any particular brand, I suggest reading a review in a good trade magazine. I use all Senco nail guns, mostly because I started out working with them and have grown accustomed to their feel. Senco is also the only

A Homemade Roller Stand

Before I built an outfeed table for my Rousseau table saw stand, I used a long roller that I also built myself *(see photograph)*. I turned the roller on my lathe from a cast-off piece of 4x4 redwood. The roller mounts on dowels secured in the shoulders of a long U-shaped support. Mortise-and-tenon joints and stout oak ensure stability and durability. The long, adjustable-height neck passes through two guides. The top guide is also a yoke — notched along its length for several inches in both directions of the center mortise — and can be squeezed tightly against the neck with one turn of a wrench. The wrench is welded to a hex bolt that passes through the slot in the neck and is secured in the opposite side by a threaded insert. I built the legs wide so that the roller stand wouldn't be a pushover.

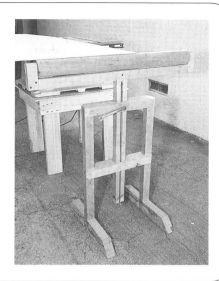

manufacturer I'm aware of that sells repair kits that are stocked by most tool stores. The Top Air and Bottom Air kits repair air leaks; the Spring Return kit repairs nail feed problems; and the Driver Kit repairs broken or damaged drivers *(Figure 3-8)*. I keep one of each type of kit in my van, for each of my guns, so that I can repair my nailers in the field and keep working.

Nailer Sizes

For a finish carpenter who's just starting out, two guns are a must: a trim nailer that's able to shoot between 1¼-in. and 2½-in. nails, and a pin nailer that can shoot small brads. I use a Senco SFN 40 trim nailer because it shoots 15-gauge finish nails from 1¼ in. to 2½ in. long (smaller gauge means thicker wire). This gun is great for hanging crown moldings, nailing paneling, setting door jambs, and applying most casings *(Figure 3-9)*.

To attach thin casings to jambs, to join corners in baseboard and crown, and to apply delicate moldings, I use an old Senco LSII that fires 18-gauge slight-head brads from ⅝ in. to 1 in. long *(Figure 3-10)*. Senco's new replacement model, the SLP 20, is a great improvement and fires both slight-head and medium-head nails, in sizes from ½ in. to 1⅝ in.

I have two other guns I rely on for special occasions *(Figure 3-11)*. I own an older Senco SJS finish stapler that shoots a staple with a narrow 3/16-in. crown in lengths from ⅜ in. to ⅝ in. Senco now manufactures a replacement model, the SJ10, which accommodates a wider range of staples, in both crown size and length. I find these tools are

Figure 3-8. *Senco nail gun repair kits are available for repairing air leaks, nail feed problems, and broken or damaged drivers. Most tool stores stock repair kits for Senco nail guns.*

superb for installing ¼-in. plywood skins, cabinet backs, and for a variety of other specialized chores.

Because a finish carpenter needs to be prepared for anything, I also own a framing gun, an old Senco SN II. I use the old framer for setting trimmers, framing window and door openings, and the general rough carpentry I'm occasionally required to perform. This gun is almost an antique. And you wouldn't want to accidentally drop it on your foot! I think it's made from cast iron because it weighs nearly 10 lbs. Newer framing guns, such as the Senco 600 and 650, weigh closer to 8 lb. and handle a wider range of fastener sizes.

Figure 3-9. *This Senco SFN 40 trim nailer shoots 15-gauge finish nails from 1 1/4- to 2 1/2-inches long.*

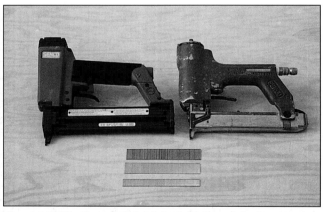

Figure 3-10. *Senco's SLP 20 (at left) fires slight-head and medium-head nails, in sizes from 1/2- to 1 5/8-inch. It replaces the old LS II (at right), which fires 18-gauge slight-head brads, from 5/8- to 1-inch. Both are good for delicate work.*

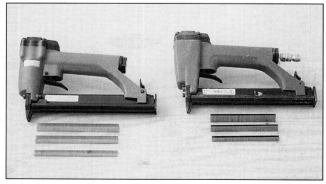

Figure 3-11. *Senco's SJ 10 stapler (at left) accommodates a wide range of staples, in both crown size and length. It replaces the old SJS finish stapler (at right). Both tools are superb for installing 1/4-in. plywood skins, cabinet backs, and for a variety of other specialized chores.*

Choosing Extension Cords

Extension cords can be the bane of a carpenter's life. The best cords are also the heaviest so they're a pain to coil and store. But using an extension cord that's too small will shorten the life of a power tool. Smaller extension cords can't satisfy the electrical thirst of some tools. (Whenever a power tool demands more electricity than the extension cord can supply, resistance occurs and resistance causes the motor to heat up and burn out prematurely.)

For running long distances, most power tool manufacturers recommend a #10 AWG (American wire gauge) cord. I find that size cord unwieldy, but I carry a #10 and roll it out when I have to set up a full shop 100 feet from a power pole. I use #12 cords for most of my work and carry several in different lengths: 100 ft., 50 ft., and 25 ft.

COMPRESSORS

There are a lot of portable compressors out there, and as with nail guns, choosing the right one for you is a matter of objective comparison and subjective taste. Even the smallest compressors, capable of delivering 120 lb. of air pressure, have enough capacity to run a finish nailer. But larger tanks are helpful for those times when two carpenters are plugged into one compressor, for occasional framing, and for those long blasts of air needed to clean dust from tools and shirt interiors at the end of a hard day.

I've gone through a few inexpensive compressors and have learned that only professional-duty motors and tanks are worth purchasing. Once again, I suggest reading a good review before purchasing a compressor, but be sure to plug in any compressor you're planning to purchase and listen to it before buying (some compressors are so loud you can't think near them, let alone converse with another carpenter; other models seem to whisper).

Pay attention to the amp draw, too. My little Rol-Air has a 3/4-h.p. motor and draws only 15 amps *(Figure 3-12)*. I've never popped a breaker with my compressor, even when running it on a #12 100-ft. extension cord (see "Choosing Extension Cords"). Another worker on my crew has a larger 20-amp, 1-h.p. compressor and it pops breakers all the time, even on a #10 extension cord. He usually has to leave his compressor at the temporary power pole and run lots of air hose.

SMALLER TOOLS

My van is loaded with power tools and not because I'm a tool junkie *(Figure 3-13)*. I don't buy a tool, let alone devote valuable van space to one, unless it earns its keep. My collection of tools spans more than thirty years and I carry a lot of specialty tools that many carpenters might never need. However, there are basic necessities that serious carpenters should always carry to every job.

Circular Saws

Although table saws and sliding miter saws have made the life of a carpenter much easier and more accurate, a lot of jobs, such as door-hanging, shelving, most framing, and countless other chores, must still be performed with circular saws.

Two types of hand-held circular saws are manufactured: worm-gear saws and direct-drive saws. Choosing the right one for you is mostly a matter of personal preference and cutting style. For beginners, my advice is to pick the one that feels the most comfortable in your hand, isn't too heavy to hold safely, and cuts along a straight line with relative ease.

I began using a worm-gear Skilsaw, and I still carry one in my truck. I use that big saw rarely now, mostly for framing and for cutting doors that are more than 1 3/4 in. thick. For the majority of my finish work, I prefer a lightweight Porter Cable 6-in. Saw Boss *(Figure 3-14)*. At first heft, my little 6-in. saw doesn't seem like a professional tool, but I've used it for several years now and can attest to its durability and power.

My worm-gear saw has a 12-amp motor and turns 4,600 rpms, while the little Saw Boss draws only 9 amps and turns 6,000 rpms. The smaller saw cuts more smoothly and with less tear out, especially using a thin-kerf carbide blade. It is also perfect for ripping and cutting long scribes on MDF shelving.

For cutting delicate scribes in casing and small moldings, I use a Makita 4200 N 4 3/8-in. circular saw. This small saw turns 11,000 rpm, and its small blade can follow almost any squiggling scribe line. While it's meant for paneling, I also use the Makita 4 3/8-in. saw to cut aluminum weatherstripping, thresholds, and sill nosings.

Figure 3-12. *A small compressor capable of delivering 120 pounds of air pressure will have enough capacity to run a finish nailer. But a large tank will make it possible to run a framing nailer or multiple finish guns. This 3/4-h.p. Rol-Air draws only 15 amps.*

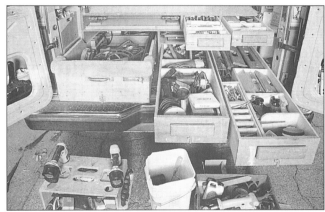

Figure 3-13. *Every tool in the van has to earn its place there. My collection of over 30 years includes many specialty tools that only finish carpenters might need.*

Figure 3-14. *A worm-gear saw (at right) is useful for cutting wood framing, as well as for trimming doors more than 1 3/4 in. thick. A lightweight circular saw, such as this 6-inch Porter Cable Saw Boss (at left), will suffice for most finish jobs.*

Figure 3-15. *A pair of 7/8-h.p. routers works well for routing hinge, latch, and strike mortises, as well as other general-purpose jobs.*

Figure 3-16. *A 1 1/2-h.p. router mounted beneath a table-saw stand is used for heavy work like dadoes, large rabbets, and molding bits.*

Figure 3-17. *A laminate trimmer with a roundover bit can be used to ease the edges of material ripped on the table saw.*

Figure 3-18. *Plunge routers do a variety of jobs, including making dadoes, rabbets, pocket mortises, and even intricate molding details.*

Routers

Of all the tools I carry, I enjoy routers the most. Fitted with the right bit or used in conjunction with a proper template, a router turns a difficult tedious chore into a relaxing easy job. For repetitive tasks — and finish carpentry often involves repetitive work — routers are especially fast and accurate.

For general-purpose routing of hinges and hardware components, I use a pair of Porter Cable 100 7/8-h.p. routers *(Figure 3-15)*. I use one for hinge mortising and the other for latch and strike mortising. For heavier work, like dados, large rabbets, and molding bits, I turn to a 1 1/2-h.p. Model 690, with a D-handle and integral trigger switch. I mount this router beneath my Rousseau table saw stand also *(Figure 3-16)*.

I still carry my very first router, an old Craftsman model I received as a gift when I was eighteen years old. But I use this antique only for light work, such as cutting kerfs with a slot cutter or running small beads with a molding bit (see "Router Bits").

Though my Bosch laminate trimmer isn't considered a true router, it fits best in this category because of its principal uses. While it works well for its intended purpose of cutting laminate flush with substrate, I always store my trimmer with a 1/8-in.

> ## Router Bits
>
> Here's another tool collection that should develop and grow over time. Though several router bits should be purchased immediately, profile, molding, and specialty bits ought to be purchased only as they're needed, both to reduce initial outlay, and to be sure of buying the right tool for the job.
>
> I've collected an assortment of router bits. Oddly enough, I've found that once I've bought a good bit, even if it seems esoteric at the time, I begin to use it frequently.
>
> Standard bits that every router owner should have include 1/2-in. straight bits for use with hinge and lock templates, as well as 1/4-in. and 3/4-in. straight bits for cutting shelving dados. I carry these bits with 1/4-in. and 1/2-in. shanks, and several extra-long ones, too, for cutting those deep mortises for Soss invisible hinges and extension flush bolts.
>
> Both a standard ogee bit and a Roman ogee bit are essential for running edge beading on molding and shelving, and these two bits can be used for rounding over bull-nosed edges, though corner round bits are also designed especially for that purpose. Classical bits with cove and bead patterns are useful for more dramatic detailing. A V-groove bit is handy when working with V-groove paneling, and for adding individual board details to wide glue-ups. I also carry core box bits in several sizes for cutting fluting.
>
> Slot cutters are useful, too, in different sizes from 1/16 in. to 1/4 in. These three- or four-wing blades are great for cutting repetitive rabbets; they're the best tool to kerf jambs for weatherstripping or bull-nose corner beading; and they work better than a biscuit jointer for joining paneling stiles and rails.
>
> Flush-cutting trimmer bits, which are guided by bearings, are also useful. I use the shorter ones, with bottom-mounted bearings, for cutting laminates perfectly flush. I use the longer bits — with top-mounted bearings — with pattern templates.

roundover bit, so that after ripping material on a table saw I can quickly ease the edges *(Figure 3-17)*. I also use my laminate trimmer with small beading bits, and I mount the trimmer in a specially designed tool handle for retrofitting existing jambs with silibead weatherstripping.

Plunge Routers

Of the five routers I carry, I enjoy my two plunge routers the most, because they save me the greatest amount of time and frustration. Many of the tasks I once performed with a circular saw, drill bits, and chisels, I now knock out with a plunge router. These include plowing wide dados, deep rabbets, pocket mortises, and even intricate molding details *(Figure 3-18)*.

The Bosch 2-h.p. 1613 EVSB is a fine tool. Equipped with an electronically controlled soft-start motor, the router is exceptionally easy to hold by hand, and the controls are sensibly arranged, especially the depth lever. Several other features are worth noting: An arbor lock makes bit changing very easy; the quick-release template guide is a pleasure for changing between template guides and bearing mounted bits; and a sensitive fine-adjustment knob can control depth of cut by the thickness of a sheet of paper.

I also have a 2-h.p. Porter Cable 97529. It's a workhorse of a tool, though I don't favor it nearly as much as I do the Bosch. The Porter Cable lurches too much when I change the depth of cut, the base isn't solid and rocks on the narrow edge of a door, the fine-adjustment depth knob isn't easy to operate, and the safety switch on the trigger is a nightmare of poor engineering. But the motor is powerful and I can push a 3/4-in. bit down the length of a door stile and plow an extension-flush-bolt dado in a hurry. In addition, the accessory fence on the Porter Cable is excellent.

Cordless Tools

I used to carry two push-type Yankee screwdrivers, and I used them frequently for driving screws and drilling small pilot holes. I still carry scars from those push-tools, from the times a slot-bit would slip off a screw head and bore a hole in my hand. Fortunately, I replaced those Yankee drivers with two cordless drills.

Cordless tools, like nail guns, are one of the technical advances that make carpentry easier and more enjoyable than it used to be. But choosing which brand to purchase isn't easy at all. I started out using a small Bosch cordless drill, then bought the Makita 9.6v long-handled driver soon after it

Figure 3-19. *A cordless saw is handy for making quick cuts, as well as for trimming drywall around window and door openings. An 18-volt cordless flashlight is easier to handle than a drop-light in closets and other windowless spaces.*

appeared on the market. Later, I upgraded to 12v tools, and now I use a combination of Makita and DeWalt tools in 14.4v and 18v sizes.

Once again, I suggest reading a review before investing in any cordless tool, and try to stay with the same manufacturer for all your cordless needs, so that the same charger can charge different batteries. Several manufacturers offer full-line systems, including Makita, Porter Cable, DeWalt, Hitachi, Panasonic, Bosch, and Ryobi.

Also, before purchasing any cordless tool, consider the volume and type of work the tool will perform. Batteries with higher voltage provide more torque and longer run time after each charge, but higher-voltage batteries also have a shorter life cycle—they can't be charged as many times as lower-voltage batteries. And higher-voltage batteries also cost a lot more to replace.

I rarely walk onto any job site without carrying two cordless drills. For the countless numbers of screws I drive nearly every day, I use a 14.4v driver drill (DeWalt DW991). For me, the size and voltage is perfect for constant heavy use. For drilling pilot holes and holes with spade bits, hole saws, and masonry bits up to $1/4$ in., I carry an 18v cordless hammer drill (DeWalt DW998) because it provides more power for boring holes in wood and percussion drilling in concrete.

I also own a cordless saw and flashlight, both of which I use frequently *(Figure 3-19)*. The saw is handy for making a few quick cuts, even in 2x framing material. Because it produces a low volume of dust, the saw is also useful for cutting drywall on window and door remodels. The 18v flashlight often saves me from carrying a drop-light around while working in closets and windowless bathrooms.

Drills

Not every job can or should be performed by a cordless tool, although sometimes it's tempting to try. After using my cordless tools extensively for a few years, I learned to use corded tools whenever possible. For instance, if I'm hanging only one door, I'll use my cordless drills. However, if I'm installing doors all day, I'll save the wear and tear on my cordless tool batteries and turn to the following plug-ins *(Figure 3-20)*.

For driving handfuls of hinge screws, I prefer a screw shooter. Screw shooters meant for drywall work vary from 2,500 rpm to as much as 4,000 rpm, but the Milwaukee 6543-1 that I use for installing hinge screws has a variable speed motor with an rpm range from 0 to only 1,000. The slow speed, together with a positive-drive clutch, makes this special-application tool perfect for installing temperamental hinge screws.

For drilling pilot holes for screws or boring jambs for lock strikes — and for all light-duty repetitive drilling — I depend on the lightweight $3/8$-in. Bosch 1020VSR drill. But for boring doors for locksets and drilling large holes in wood and masonry, I like to use a heavy Bosch 1194VSR — it's a two-speed, variable-speed, $1/2$-in. hammer drill, and it's powerful and durable.

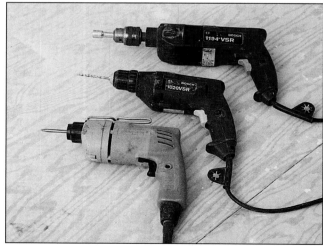

Figure 3-20. *For an intensive round of door hanging, it's better to save your cordless tools and turn to the plug-in drills. I use the Milwaukee screw shooter (at bottom), as well as a Bosch 1020VSR (at center) and a Bosch 1194VSR (at top).*

Figure 3-21. *A reciprocating saw (at top) is a remodeler's right hand. A jigsaw (at bottom) is the tool for coping large moldings and for cutting circles and blind notches.*

Reciprocating Saws

Remodeling is often the first type of work carpenters do when they start out in the trades, and there's no tool more popular on a remodel than a reciprocating saw. I bought a reciprocating saw right after my first circular saw, and even as a finish carpenter, I still use one often. For cutting door and window openings, for removing old jambs, and for generally tearing things apart before exercising the fine art of finish carpentry, reciprocating saws are life-savers. I've gone through several of these tools, though my Bosch "Panther" model 1632VS has lasted the longest. With a variable-speed motor and an adjustable orbital cutting action, this saw goes through wood and metal quickly.

Jigsaws

Jigsaws are also required tools for finish carpenters *(Figure 3-21)*. I use mine for a wide assortment of challenging jobs, from cutting lockset holes in metal doors to fancy scrollwork on wooden corbels. I use a jigsaw for cutting aluminum, brass, and stainless-steel thresholds, as well as panic hardware rods, towel bars, and chrome closet poles. A jigsaw is the only tool for making cut-outs in cabinets and countertops, and for cutting tight scribe lines in boards that will fit around irregular stone and brick.

Jigsaws are also the best choice for coping large moldings, especially hardwood crown molding and baseboard. For coping, I attach a Collins Coping Foot® (Appendix 1) to the bottom of my Bosch 1587VS. Before I bought the coping foot, I used to hold my jigsaw away from the edge of the material, which isn't easy to do. The movement of the blade tends to slam the base up and down against the wood and mar the edge and face of the molding, especially when cutting tight turns and particularly on hardwood. The coping foot is one of those simple inventions that attaches easily and changes a difficult, sometimes frustrating job into a pleasant and accurate experience (see Chapter 8, "Baseboard," page 120).

Sanders

For finish carpenters and cabinetmakers, belt sanders and palm sanders are tools that see daily use *(Figure 3-22)*. My belt sander saves me countless hours of labor when I have to smooth joined boards, and I use one every time I plane the edge of a door to remove minor rough spots. Unfortunately, the learning curve for belt sanders is steep, as first-time users invariably sand surfaces unevenly, leaving dips, valleys, and ridges in joined boards. My best advice is to use a machine that has a large sanding surface. I consider the 3x24-in. belt on my Makita sander to be a minimum size: Smaller belt sanders are too difficult to control.

For sanding out the marks left by the belt sander, and for generally applying the finishing touch needed for fine woodwork, I use an orbital palm sander. I've found that the Porter Cable 335 palm sander isn't too aggressive and doesn't leave those small orbital marks or scratches — even with 80-grit sandpaper — commonly caused by some pad sanders. Yet the 335 makes quick work of what is otherwise a

Figure 3-22. *The 3x24-inch belt on this Makita sander (at right) is the minimum practical size; smaller belts are just too hard to control. The Porter Cable random-orbit sander (at left) removes the marks left by the belt sander, and is great for applying the finishing touch on fine woodwork.*

Figure 3-23. *A full set of levels is needed for setting windows and doors, as well as a variety of trim and hardware.*

Figure 3-24. *A builder's level (at left) or laser (at right) shoots control lines for window and door jambs and casings. The laser also makes short work of repetitive jobs, such as laying out shelves.*

nasty chore, especially when sanding face frames and flat casing miters, where sanding must be done simultaneously with the grain and across the grain.

Levels

Nearly every carpenter's first level is 48 in. long. Most carpenters get by with that single level for years, sometimes with the exact same tool (at least many of the levels I've seen on job sites seem to be that old). Some craftsmen invest in a 24-in. level, too, for setting door jambs, closet shelving, towel bars, and an assortment of short things on short walls. But I've learned that a full set of levels, in various lengths, is one of the most important tools for a finish carpenter to own *(Figure 3-23)*.

Investing in a full set of levels is extremely expensive. Without question, a good short level attached to a perfect 78-in. straightedge can give you the same accuracy as a 6-ft. 6-in. jamb-setting level. And perfect straightedges can be fashioned for other jamb sizes, too, including large openings.

But for my work, which involves setting a great many windows and doors, and installing a wide variety of trim and hardware, a full set of fine levels is especially useful. I've accumulated and frequently use the following sizes: a short 12-in. torpedo, a two-footer, a 32-in. jamb-head level, an older 48-in. model, a newer 5-ft. model, a 6-ft. 6-in. jamb-leg level, and a long 8-ft. level for tall jambs and wide windows and doors.

Levels must always be accurate and dependable. Accurate levels are not only a pleasure to use and protect, they assure precise work the first time. (I once used adjustable levels for all my work, until I had to reset a house-full of door jambs because the single horizontal bubble in my 2-foot level wasn't adjusted correctly.) Levels must also be durable and able to withstand the foolishness of green helpers and the rush of other tradespeople.

Builder's Levels and Lasers

I don't lay out houses or additions anymore, and I haven't poured concrete in more than a decade, but I still use my old builder's level for shooting control lines in large custom homes *(Figure 3-24)*. Control lines assure that window and door jambs will align perfectly, that door and window tops will flush out exactly, and that head casings on windows and doors will form a straight and true line.

Lasers are one of the latest technical advances in the building industry. For several years I've seen them used by T-bar ceiling crews, surveyors, and foundation contractors. Recently I used a laser to lay out a coffered ceiling so that the electrician and plumbers could install flush lights and fire sprinklers. The bright red line projecting across the ceiling, down the wall, and back across the floor saved countless trips up and down a ladder and precious hours I would have spent waiting for a co-worker to settle a dancing plumb bob! And because lasers set up so fast, I've found myself using the red dot for jobs that would have once meant dragging a level from wall to wall — like laying out shelving lines in large closets or pantries, or wainscoting and library paneling.

In the Tool Belt

They serve the same purpose whether they're called tool belts, nail bags, tool bags, or nail pouches. Unfortunately, some nail bags and tool belts tend to collect a lot of trash and sawdust. I've seen carpenters empty tool bags full of mixed up nails, screws, odd pieces of hardware from last week, broken pencils, frozen pliers, dull utility knives, and chipped chisels. But tool belts should be treated with respect: They should be organized carefully because the tools you carry in your belt tend to be the most important and most commonly used instruments of the trade.

The working habits of carpenters vary from one personality to the next, and the contents of each carpenter's tool belt often illustrate that difference. Yet craftsmen working on opposite ends of the country often carry remarkably similar tools in their belts.

Besides a 25-ft. tape measure and a lightweight Stiletto titanium hammer (Appendix 1), I carry the following in my bags:

- A small prybar or lifter, which I reach for repeatedly every day
- A pair of offset wire cutters, or dikes, for cutting or pulling nails

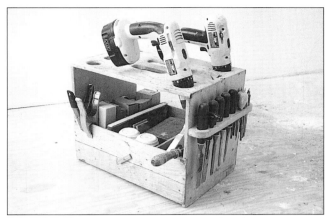

Figure 3-25. *I carry my most-often-used tools in a homemade tool box that serves also as a work bench, step stool, and seat.*

- Two nail sets: a #1 and a #2
- A cold chisel for bending hinges, driving up stubborn hinge pins, chipping bits of concrete from beneath a sill, etc.
- A utility knife with spare blades
- A pair of scribes (General Tool Co.) fitted with a mechanical pencil (Cross Pen)
- A $2^{5}/_{10}$ medium pencil (hard enough to leave a sharp line, but not a scratch)
- A square, with a small bubble and a $1^{1}/_{2}$-in. blade
- A small crescent wrench, for bending hinges and changing circular-saw blades
- Three or four small drill bits, from $1/_8$ in. to $1/_4$ in., for drilling pilot holes
- A tapered countersink, for drywall screws
- A carbide reaming/grinder bit, for adjusting lock strikes
- Two Vix bits (#3 and #5), for drilling pilot holes
- A couple dozen #4 finish nails, for stringing jambs or tacking the occasional casing or door stop
- A small mixture of gold drywall screws, in $1^{1}/_{4}$-in., $1^{5}/_{8}$-in., and 2-in. lengths
- A four-bit screwdriver

The Tool Bench

For many years I carried a crippling amount of weight in my tool bags. And like most guys out there, I made far too many trips going back and forth to my truck for items that either I'd forgotten or wouldn't fit in my bags. I hated the lost time and the lost energy on those trips back and forth. I finally came up with the idea of a handmade tool box, which also serves as a work bench, step stool, and seat *(Figure 3-25)*. I transferred many of the tools I

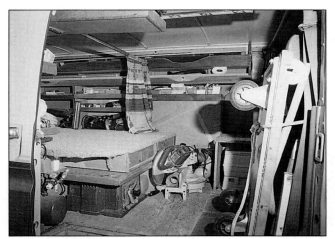

Figure 3-26. *When laying out space in a van or truck, start with the largest items — then use the remaining large spaces for as many drawers as possible, along with shelves and cubbies.*

Figure 3-27. *A storage system that's built in to the back of a pickup truck must be covered with a camper shell. Build the cubbies and drawers tall enough that your large tools store below the upper wooden shelf, leaving it free and clear for sheet goods, doors, and trim.*

commonly use to my "tool bench," but to keep it from becoming overloaded, the only items I allow to remain on board are those I use nearly every week. (See "A Homemade Tool Bench.")

FITTING IT ALL IN: TRUCKS AND VANS

Choosing between a truck, a truck and trailer, or a van is a personal decision and depends on driving comfort, weather and climate, as well as work habits and job types. I first drove a full-size pickup, then switched to a king-cab import that got better mileage. About six years ago I bought a van and now I'll never go back to trucks. I don't do that much commercial work anymore, so I don't have to worry about my tall 1-ton van fitting into underground parking garages. The van also allows me to carry all my tools all the time, in addition to a considerable amount of material, both inside the van and on the racks above.

Whether you like to drive a truck or a van, or haul a trailer with all your gear, organization is always the key. For finish carpenters, it's important to carry all the tools I've previously described, and still have room for jambs, trim, doors, windows, and hardware.

I've used side boxes, cross boxes, low roll-out drawers, and a combination of all three to organize my gear, but the best method is a series of shelves, drawers, and cubbies, built into the back of a van or pickup truck *(Figure 3-26)*. For this system, the pickup truck bed must be covered with a camper shell, though a large one isn't necessary *(Figure 3-27)*.

If you're about to build a storage system for your tools, start by laying out space for the largest items — door bench, table saw stand, chop saw stand, ladder, chop saw, table saw, etc. — then use the remaining large spaces for as many drawers as possible. Don't make the drawers too big or they won't slide when you fill them will tools. Make them just large enough to store your tools in separate spaces, secured by dividers. Any small areas that remain can be used for long narrow nail and screw bins, hardware storage, level storage, etc.

Build the cubbies and drawers tall enough so that your table saw, chop saw, and compressor will store beneath the top wooden shelf, because sheet goods, doors, jambs, and molding need to be loaded easily and flat, without any tools in the way. Loading and unloading a pickup truck is performed mostly from the back, so allow enough room for a table saw, chop saw, and compressor at the rear of the bed.

Storage space is even more accessible in a van *(Figure 3-28)*. My compressor, table saw, and chop saw ride just inside the sliding side door, but I built my drawers and cubbies short of the rear door, so that my work bench would fit in the back, along with a milk crate and trash bucket.

Shelves on the sides of my van hold my hinge templates, levels, weatherstripping, miscellaneous hardware, and a two-sided, 3-foot ladder.

A Homemade Tool Bench

Some of the handiest things I use as a finish carpenter are homemade, like custom router templates, a chop-saw stand, a door bench, and especially my tool bench. I never leave my van for a job site unless I'm carrying my tool belt and my small tool bench — usually with a cordless drill or two stuck in the top.

The footprint of my bench was easy to decide upon: I made it small enough to fit inside a milk crate, so that I could stack both inside my van, or stack the two for a taller step stool on the job site.

The height of the bench was determined by several factors: I needed room for a small drawer beneath the lower shelf, to hold an assortment of wrenches, bits, and odd tools; I needed to safely store a set of long chisels at the front of the bench; and I wanted my cordless drills to fit inside the top of the bench, without hitting or interfering with the contents inside. Also, I'm 5 ft. 7 in. tall and felt that another 14 inches would put me at eye-level with the top of a 6-ft. 8-in. door. Finally, I should also admit that I gave serious consideration to the perfect height for a working seat — 14 in. seemed just right.

I made the bench from 1/2-in. Baltic birch plywood, and rabbeted the corners for strength and durability. The side legs run an inch past the bottom of the bench, so they'll clear debris on the floor and provide a secure platform. I mortised in the narrow 1/4-in. plywood sides, to add shear strength and also eliminate corners that might catch and splinter.

I drilled four 2-in. holes into the top of the bench to hold drills, and I used a crude template to rout in two rectangular dishes in the top so I would have a place to put small screws and fasteners without having them fall off the top and get lost.

I carry a complete set of chisels and spade or paddle bits. These are attached to the front of the box where I can easily reach them while sitting before a door jamb and installing hardware. This collection includes an awl and a 1/8-in. T-handled Allen wrench, for Baldwin-type hardware.

I installed a few dividers inside the bench to separate and help organize boxes of screws, plastic plugs, wood fillers, glue, silicone spray, etc. And finally, I added a low drawer to store frequently used wrenches, pliers, marking tools, masonry bits, driver bits, etc. Inside the bench I store the following:

- A drill index
- Boxes of drywall screws, mostly in gold and ranging in size from 3/4 in. to 3 in.
- A small tub of spackle and one of latex wood filler
- A large shim for holding doors open or up
- A plastic angle finder for finding unusual miter angles
- Several masonry bits and an assortment of plastic anchors for installing thresholds on concrete floors
- A small bottle of glue
- A can of silicone spray
- A sharp dovetail saw
- About 30 ft. of fly reel backing, for cross-stringing jambs and checking long sills and jamb heads
- A sanding block
- A reveal jig
- A roll of blue masking tape

This collection of fasteners, fixtures, and miscellaneous stuff might seem too bulky and heavy to carry every day. But each time I decide not to carry the full kit, I seem to spend more time walking back and forth to my van than actually working.

A wooden box loaded with oft-needed tools and supplies will reduce trips back and for to the truck. To minimize weight, the box should include only tools that are used at least once per week. The bench shown is small enough to fit inside a milk crate.

Figure 3-28. *With good organization, I can carry in my van everything I need most often: a compressor, table saw, and chop saw; a workbench, milk crate, and trash bucket; drawers, cubbies, and shelves for tools, and hardware. And there's plenty of room for materials on top of the shelving.*

MISCELLANEOUS SUPPLIES

Like those of many other tradespeople, my truck is like a traveling storage bin. To avoid quitting work early or making sudden unexpected trips to the hardware store in the middle of a job, I carry bulk amounts of the following frequently used supplies and equipment:

Screws. A wide assortment of sizes and types, from drywall to self-taping, and from coarse to fine thread.

Caulk. In addition to multiple tubes of both Brilliant White and Clear Silicone Caulking, I carry several colors of latex caulk, a few tubes of Big Stretch, a few tubes of White Bathroom Silicone, a few tubes of liquid nails or construction adhesive, along with an assortment of polyurethane caulk for moldings and waterproofing.

Sandpaper. Because helpers never have what they need, I stock sandpaper in bulk packages, and in several types, from 80 grit to 220.

Drop cloths and plastic sheeting. Clients like to have work done on their homes, but they hate sloppy workmanship. I always carry enough tarps and plastic to protect two rooms, and enough to create a wall or envelope, too.

Ladders. In addition to my 3-foot aluminum ladder with steps on both sides (which I depend on for almost all my ceiling work and 8-foot door installations), I also carry 6-foot and 8-foot aluminum ladders, so that I can reach tall ceilings or build a quick scaffold between them.

Rope. Rope isn't just for tying down loads. I often use a 100-ft. piece of $1/2$-in. braided nylon for hauling material up the side of a multi-story building.

CHAPTER 4

Exterior Doors

Before any moldings can be applied in a home, the windows and doors (or at least the jambs for the windows and doors) must be installed. And because setting doors and windows is the first order of business for most finish carpenters, that's where the real action in this book begins.

Door and window installation is easier and less time-consuming if everything is ordered correctly. In addition, interior trim carpentry goes much faster if the windows and doors are ordered accurately and set properly.

In this chapter, I'll first review the system I use to ensure that door orders produce the doors they're supposed to. Then, I'll describe step-by-step procedures for installing exterior doors (both sliders and swinging doors) that ensure smooth, accurate, and profitable finish work. I'll cover windows in Chapter 5 and interior doors in Chapter 6.

ORDERING DOORS

As a finish carpenter and contractor, I often work as a subcontractor. Though I supply material as well as labor for many of the jobs I do (allowing me to control both the delivery and the correct purchase of doors, windows, and trim), I also install material supplied by general contractors and clients. I've made and witnessed many mistakes on door orders, and I've learned that a significant amount of time and money can be saved if a routine form is used (see "Exterior Door Takeoff Form," Appendix 2).

Routine forms are the best insurance against costly mistakes, and filling out each category carefully is the first requirement for a successful and profitable job. The takeoff form I use for window and door orders can be recreated easily in any spreadsheet program. A spreadsheet form organizes information, prints clearly, and calculates costs automatically. When I print my takeoff form for other tradespeople, homeowners, or general contractors, I leave off the cost figures. But the other categories include essential information for all parties; I spend whatever time it takes to complete these categories carefully and accurately.

Walk the Plans

Before I begin any takeoff, I "walk the plans." From the front door I usually move to the right and find each exterior opening on the floor plan. I use colored highlighters to mark each window and door, with one color for doors and another for windows.

I also walk the elevations and check that the openings and details between the elevations and the floor plans match. Sometimes windows and doors are hidden in upper elevations that are difficult to see on a floor plan. On the elevations I mark the plan number for each exterior opening, so that I can later compare the number of doors and windows on the floor plan to the number on the elevations.

I walk the framing plans as well, to learn where sheathing or shear paneling is specified and to determine its thickness. Then I turn back to the floor plan and use a red highlighter to note shear-wall locations and sizes beside the affected walls.

A Rough Take-Off

Because it's difficult to work on a large set of plans and type directly into a computer, I usually begin with a rough list. I write down each and every

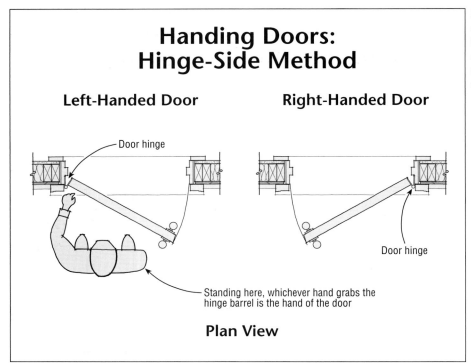

Figure 4-1. *Using the hinge-side method, face a door that's swinging towards you. If the hinges are on the left, it's a left-handed door; if the hinges are on the right, it's a right-handed door.*

opening, along with its location, on a legal pad. While it's tempting to find and combine like-size door units, errors are easily made, and compiling similar-sized openings should be avoided at this stage: One 5/0 x 6/8 pair of French doors at the front of a home might have beveled glass, while another at the breakfast room might be tinted. The jamb widths might vary, too, according to shear-wall locations and sizes. Therefore, before judging that doors can be ordered in duplicate, first fill out the computerized order form thoroughly. Then, only once it's certain that like-sized units are exact duplicates, adjust the quantity amount.

Begin by noting the dimensions of each door unit, writing first the width and then the height. Next, find and record the necessary jamb width. Doorjambs must be wide enough to accommodate the stud size, the shear wall, and the drywall, so that casing can be applied without difficulty after drywall. The jamb width should not include the thickness of the exterior trim, though the type and size of the exterior trim should be included on the form.

Of course, all doors are handed, too, so the handing must be noted, as well. Two different methods are used for handing doors; therefore, be sure to tell your supplier which method you're using.

Hinge-side door handing. Some door companies use the hinge-side method, in which case you should imagine that you're standing with the door swinging toward you. If the hinges are on the right, it's a right-handed door. If the hinges are on the left, it's a left-handed door. A sure way of practicing this method is to face the door and grab the barrel of the hinge. The hand that grabs the hinge is the hand of the door *(Figure 4-1)*.

Butt-in-the-butts door handing. Some door manufacturers use the butt-in-the-butts method, in which case you should imagine that you're standing with your back against the hinges (butts), the door opening away from you. Then the direction of the swing is the handing of the door: If the door swings to your left side, it's a left-handed door *(Figure 4-2)*.

As you can see, these two methods result in opposite handing, so be sure to clearly communicate which handing method you used. And remember, handing applies to pairs of doors, too: one door is active and receives a lockset; the other is inactive and gets a dummy handle and slide bolts.

Other takeoff items. For exterior doors, many manufacturers also need to know what type of threshold or sill is required, and if there is any special hardware preparation, including flush bolts and dead bolts.

Unlike with windows, all glass in doors must be tempered, but other glazing requirements can vary dramatically. Specify whether the doors are dual-glazed or single-glazed, tinted or clear. Stipulate

Figure 4-2. *In this method, stand with your back against the hinge side of the opening. If the door swings to the left, it's left-handed; to the right, it's right-handed. Since the two common handing methods yield the opposite results, be sure to specify which method you are using.*

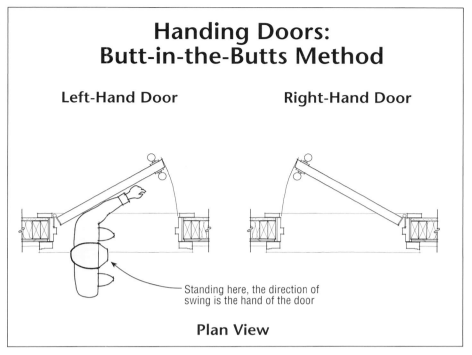

also if the doors are true divided-lite (TDL), single-lite (one large pane of glass), or have lites divided by grids. The lites in French doors should match the cut-up layout in adjacent windows, so I also note any special divided-lite layouts.

Finally, I include the rough opening size for the framers and the contractor.

INSTALLING DOOR FRAMES

Not long ago, I answered a frustrated call from the wife of a do-it-yourself homeowner. She was having a large party that weekend and her sliding door (which her husband had installed) needed to be "adjusted." I arrived at the home and knocked on the front door carrying my wooden toolbox and a cordless drill. I figured it would be an easy fix.

With great drama and percussion, the homeowner showed me that she could lock the door only by slamming it up against the jamb. I could barely unlock it. I held the unlocked door up close to the jamb and recognized part of the problem: the door touched the jamb at the top but was 1/4 in. away from the jamb at the bottom. Hey, this'll be a snap, I thought. I'll just raise the front wheel on the door and the lady will think I'm a genius!

But when I slid the door back, to get at the wheel adjustment screw, the slider would open only halfway. The back of the door rubbed so badly on the head of the jamb that if left uncorrected it would eventually wear a groove in the jamb.

Hey, this'll be a snap, I thought. I'll just lower the back wheel and the lady will think I'm a genius! But the rear wheels were already drawn all the way up inside the door and the back of the slider nearly rubbed on the sill.

The problem was more complicated than I had thought; the situation required a thorough examination. I stepped back from the opening and instantly saw a 3/4-in. bow in the wooden sill. The oak sill rose toward the center of the opening like a gently graded hill. I looked up and noticed that the head jamb mirrored the sill, with an equal bow.

This might not be easy to fix, I thought, and bent down on my knees for a closer look under the sill. I pulled the carpet back at the center of the sill, expecting to see a hump in the concrete slab. But the sill wasn't even touching the concrete.

I went straight out to my truck and fitted a reciprocating saw with a long metal-cutting blade. It was a struggle, but I finally got the tip of that anchor bolt out of there. Then I drilled three holes for screws through the wood sill. (I first used a 3/8-in. bit to counter-bore for oak plugs). I turned to a TapCon® masonry bit to drill anchor holes in the concrete. Three TapCon® screws drew that sill right down to the slab. You can imagine how easy it was to adjust the doors once I'd straightened the sill. That lady thought I was a genius.

This was a classic case of an improperly set sill, the most common error made while installing any door unit. Because they both depend on frames,

Figure 4-3. *When setting a new frame, leveling the sill is the most important step. Use the longest level that will fit in the opening, and shim the level until the bubble is perfectly centered.*

manufactured sliding doors and pre-fit swinging doors share the same installation techniques. In this section, I'll first describe effective systems for setting exterior door frames. (Because I use the same techniques to install both sliding and swinging doors, I've assembled photographs from a variety of installations to help guide this discussion.) Afterwards, I'll highlight special issues regarding sliding doors, and then for swinging doors.

Check the Opening

Avoid making costly or embarrassing mistakes during door installations by following a few simple steps:

On a remodel, before demolishing an existing door, mock up the side jambs, head, and sill of the new unit to determine exactly what the width and height of the rough opening must be. More than once I've had to return new doors because they were ordered or shipped incorrectly. This is an error best discovered before tearing out someone's existing doors.

While checking the size of the rough opening, lay a long level across the sill. If the existing rough sill isn't level, then the rough opening must be tall enough to accommodate the shims necessary to level the new sill.

Also, place the level against the trimmers or existing doorjambs, and be certain there's enough room to adjust the trimmers plumb and still fit the unit into the opening.

On new construction, it's easy to measure the rough opening, but on a remodel the height of the header is sometimes difficult to determine. If the height of the new frame seems too close to the height of the existing frame, then remove a few screws from the head of the old unit and use an awl or a stiff piece of wire to see how high the header is above the existing head jamb. Usually there's another 1/2 in. or more of clearance. In some cases, headers are padded down to accommodate door units that are only 80 in. tall. That padding or furring can be a lifesaver, but unfortunately there's no way of knowing if the header has been padded without removing the jamb or opening up a portion of the wall. On some jobs, where the fit of the new unit is in real question, I open up the wall just enough to know the truth, before demolishing the old doors and frame.

Back-Prime Wooden Sills

Most door units with wooden sills are back-primed by the manufacturer, though some aren't. If the sill isn't properly sealed, it will warp and twist. I start back-priming wooden sills as soon as I'm sure the unit is the right size. I usually apply two good coats, sometimes three, of a fast-drying exterior primer, before it's time to assemble and install the frame permanently. I check that the bottoms of the jamb legs are primed too, or the end grain will soak up moisture like a sponge.

Level the Rough Sill

While the primer on the sill is drying, I scrape the concrete slab or wood subfloor clean of old caulk and carpet nails, then lay a level across the opening. The longer the opening is, the longer the level *(Figure 4-3)*. A good short level fastened to a straight board also works. I shim the low end of the level until the bubbles are centered, and then I place additional shims between the floor and the level about every 12 in. If a rough sill is badly out of level and a lot of shims have to be used, it's sometimes necessary to use fast-drying concrete, to level the floor. On a concrete slab I secure the shims with a dollop of adhesive caulk; on a wood floor I nail the shims down so that they won't move when I slide the frame in and out during my dry run *(Figure 4-4)*.

Shim the Trimmers Plumb

As soon as I've finished shimming the floor, I double-check the height of the header to be sure the

Figure 4-4. *On a wood subfloor, tack the shims to the floor; on a concrete slab, secure the shims in dollops of adhesive caulk.*

Figure 4-5. *If the rough opening is framed correctly —1/2 in. wider than the door frame — then plumb the trimmer on the hinge-side of the jamb by tacking the shim to the trimmer up near the header.*

new unit will still fit. Next, I measure the width of the opening and check the trimmers for plumb. If the opening is much too wide or severely out of plumb, I add blocks and backing to the trimmers and shim the backing plumb.

If the size of the opening is close to the size of the frame ("close" meaning that only a few shims are needed), then I nail shims to the hinge-side trimmer to ensure that it's plumb *(Figure 4-5)*. I prefer plumb rough openings that are 1/8 in. wider than the outside dimension of the new frame. That way, the frame can be installed with minimal slop, and only minor shimming is required. So my next step is to measure from the shim on the hinge-side trimmer across to the strike-side trimmer *(Figure 4-6)*, and fasten a shim near the top of the trimmer to fill in any additional unneeded space. Finally, I repeat the same procedure at the bottom of the opening, and install whatever size shim is necessary to make the opening 1/8 in. wider than the frame *(Figure 4-7)*.

Figure 4-6. *Shim the strike-side trimmer, too, so that the space between the shims is only 1/8 in. wider than the door frame.*

Install a Sill Pan

Once the opening is correctly prepared, I cut and fit a sill pan of adhesive-backed bituminous membrane (see "Waterproofing," next page). The pan must extend inside the house far enough to create a small dam or separation between the sill and any floor covering, especially hardwood or composition underlayment. The sill pan must also extend outside far enough to lap over the building paper or housewrap by at least 6 in. Finally, the sill should climb up the trimmers about 8 in.

I begin installing the pan by fastening a patch of bituminous membrane *(Figure 4-8)* in each corner (see "Caulking Doors and Windows," page 49).

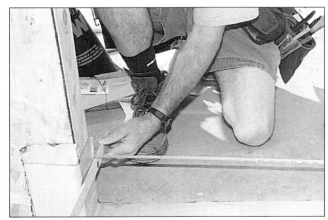

Figure 4-7. *Use a tape measure to check the bottom of the opening, too, and tack shims to the strike-side trimmer so that the jamb will just fit in the opening.*

Waterproofing

Proper window and door installation is dependent upon accurate levels, careful shimming, and precise measurement, but the first concern is waterproofing. Moisture barrier material has changed a lot over the years I've been in the business. At one time strips of 20# felt paper were cut and applied as flashing around window and door units. Sisalkraft® paper was an improvement over building paper because Sisalkraft® was fiber-reinforced and more difficult to tear. But Sisalkraft paper turned brittle with age, especially if left out in the sun and weather, which happens to most homes before they're wrapped and sealed. I've removed old windows and doors on remodels and found the Sisalkraft® flashing in tatters.

Bituthene and other self-adhesive self-sealing products — including Forti-Flash® and Ice & Water Shield® — are recent additions to the market (see photo). Bituminous rubberized asphalt flashings guarantee good results, but they can break a budget. I use self-adhering flashings only on trouble spots, such as sill pans, homes without overhangs, or mission-style stucco with bull-nosed door and window returns.

On most jobs I use Moistop®, a fiberglass reinforced paper core covered by black polyethylene on both sides (see Appendix 1). Moistop® doesn't dry out in the sun like Sisalkraft®, and Moistop® comes on long rolls, in 6-in., 9-in., and 12-in. widths. This inexpensive flashing is ubiquitous in the industry, especially for sealing vinyl, aluminum, and wooden windows.

A word of caution about combining bituminous membranes with plastic housewraps: Solvent-based products, such as rubberized asphalt membranes, have a negative effect on plastic products in

Waterproof Membranes and flashing: Bituminous membrane (left); 12 in. Moistop® (right); 6 in. Moistop® (center).

that eventually the plastic will deteriorate. If you need to install a bituminous membrane sill pan, prevent the interaction of these materials by caulking in a layer of Moistop® or paper flashing between the membrane and the housewrap.

Figure 4-8.
To prevent costly leaks and callbacks, always install a sill pan. First apply a patch of self-adhesive waterproof membrane into each corner, and then spread a longer piece across the subfloor and 6 in. up each trimmer.

These patches are the best way to solve the problem of slicing the pan at the corners. Next, I stretch a continuous piece of membrane across the opening and up each trimmer leg. Then I slice each end and fold the legs up and over the trimmers. The cut pieces adhere to the corner patches and guarantee a complete seal.

Assemble the Frame

Sometimes aluminum and vinyl door frames can be confusing. I avoid confusion when assembling a frame by finding the sill first. Sills always slope, and they always slope outside. I lay the sill down with the outside edge of the sill up.

The head jamb is often easy to recognize, too, because it's the same length as the sill. If I'm installing an 8/0 x 8/0 sliding-door unit, where all the pieces are nearly the same length, I look for the jamb with U-shaped tracks that capture and secure the sliding glass door and the screen door. Head jambs on wooden door frames are also easy to rec-

Caulking Doors and Windows

At one time I used silicone caulk to install windows and doors, but I've learned the hard way that silicone doesn't always stick. Silicone caulk must be installed in a dry, clean, dust-free, and temperate environment. Those conditions never exist on a job site, so silicone caulk rarely adheres well. Acrylic latex caulk is an improvement over pure silicone — especially acrylic latex caulk with added silicone — though these products have only minimal elasticity and often turn brittle with age. Acrylic latex caulk can't keep up with the seasonal movement of windows and doors.

The best current caulking products are urethane-based *(see photo)*, because urethane adheres well to anything — vinyl, wood, concrete, and also the skin on your fingers. And urethane caulk remains flexible for years. But stay away from inexpensive off-brands because they may have a high solvent content.

One final word about caulk: Don't count on it to last forever. Eventually caulk fails. Instead, rely on proper flashing and waterproof-shedding techniques. Use caulk only as a supplemental component.

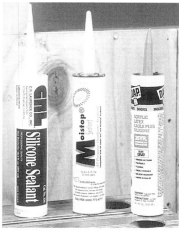

Silicone Caulk (at left) rarely adheres well on the job site. Siliconized acrylic latex (at right) is an improvement, but polyurethane caulk (at center) is the best choice.

---

ognize because they have tracks, removable doorstop, and the exterior trim is mitered twice, once for each corner.

I place each piece on the ground, and then start with the sill and the first side jamb, matching the two pieces — always with the outside edges up. I can usually determine the outside of the side jamb by the nail-fin, or by locating the milled pocket for the stationary door. The outside edge of a sill normally extends beyond the side jambs, so water will drip off away from the house. With all these clues, it doesn't take long to tell if I've got the side jamb upside down. It either fits right or it doesn't.

Most installation instructions call for caulking all the corners before assembly, which is a good practice to follow. I apply an acrylic latex siliconized adhesive in all joints before I screw the pieces together. This type of caulk combines the best properties of silicone and acrylic latex: it can be painted, has good adhesion and flexibility, and — not incidentally — it can be cleaned off easily. Urethane caulk is more durable and reliable than acrylic latex caulk, but excess urethane caulk is not easy to remove from finished surfaces, so I try to use it only where it will stay put and out of my way.

A Dry-Fit Is Crucial

Before permanently installing the door unit, I slide the frame into the opening just to be sure it's going to fit. I call this the "dry fit" because I don't apply any caulk to the rough sill, at least not yet. If screws

Figure 4-9. *Dry-fit first: To locate holes for concrete anchors, run a masonry bit through the wood sill or through the screw holes in an aluminum sill — before laying down any caulk.*

and anchors are being used to secure the sill to a concrete slab, as is usually the case on aluminum and vinyl sliding door units, a dry run is a necessity — it's the only way to locate and install concrete anchors.

With the frame in the opening, I can check that it fits properly, that it's centered in the opening, and that there's ample room to straighten and perfectly plumb the legs. If I'm installing a unit on concrete, I place a piece of tape at both ends of the sill, so that I can relocate the sill exactly in the same spot after installing the anchors. Then I run a masonry bit through the screw holes in the sill *(Figure 4-9)*. This is the best method to ensure that

Figure 4-10. *To stop water from migrating into the house, apply a bead of urethane caulk on the inside of the sill pan. Don't apply caulk on the outside of the pan because it will prevent water from draining.*

Figure 4-11. *Set the frame into the opening carefully, so the caulk stays where it's meant to be.*

all the anchors will be located precisely. (I locate concrete anchors the same way when I install wood sills, but first I use a 3/8-in. bit to counterbore for wood plugs, which I install later on top of the screw heads.)

Flash the Jamb

After all the anchors are located, I remove the frame from the opening, sweep out any dust, and insert plastic anchors. If the frame has nail-fins or stucco/brick molding, then the jamb flashing must go on next. On single-story homes with overhangs, I use 6-in.-wide Moistop®, and staple it flush with the face of the trimmers or jack studs. On homes without overhangs, or on second-story walls, I use more expensive 12-in.-wide Moistop®, and wrap the flashing inside the trimmers so that any moisture penetrating the doorjambs will drain down into the pan, where it will then be directed outside (see Chapter 5 for a more complete description of window and door flashing). The jamb flashing must extend about 12 in. above the header, where it will be covered by the head flashing after the unit is installed.

Before placing the frame permanently in the opening, I lay down a bead of urethane caulk. I locate that bead carefully at the back of the opening, in line with the inside of the wall, so that it will act as a dam and also help secure the unit to the floor. I also hold that bead to the inside line of the wall so that it won't interfere with water draining out the front of the pan *(Figure 4-10)*.

If the frame has a nail-fin or exterior molding, I run a heavy bead of urethane caulk on the back of the nailing fin, on the stucco/brick molding, or on top of the flashing. I keep that bead of caulk away from the front edge of any wood trim, or it might squeeze out when I press the unit against the finished wall — gobs of urethane caulk are difficult to remove. Of course, if there's no nailing fin or exterior trim, I just set the frame in the opening on top of the caulk *(Figure 4-11)*.

Check the Frame

I prefer to set doors (and windows) myself — rather than rely on the framers — because it makes the rest of my work easier and better looking, in at least two ways: First, I can ensure that the edge of the frame extends evenly inside the house, so that drywall won't be proud of the jamb, and applying the casing won't be troublesome. Second, I can make sure that the heads line up with adjacent heads and that the legs are parallel with walls and other doors or windows, again, alleviating problems when I apply the casing. So, if the frame has a flange (a nailing fin or exterior molding) and the jamb is the correct width for the wall, tack the upper corners in place and pull the flange tight against the exterior of the wall *(Figure 4-12)*.

If there is no flange and the jamb is the correct width for the wall, I hold the interior of the frame

Figure 4-12. *Tack the upper corners of the jamb to the trimmers, and then pull the frame in until the nail-fin or the exterior molding hits the wall — or until the frame projects inside the wall the thickness of the drywall.*

Figure 4-13. *On long sills, use a string to check that the sill is perfectly straight.*

far enough inside the wall to allow for drywall, and then tack it in position.

If the frame hasn't been ordered the correct width, then all bets are off and I set the frame tight or flush against the outside of the wall. I make sure that the frame is even with the interior of the wall because a jamb extension will have to be applied (see Chapter 7, "Casing"). Then I tack it in place.

Straight Sills Are Essential

On a slider or long frame, I adjust the sill to the tape marks made during my dry fit, and then secure each leg temporarily with a screw or half-driven nail through the face of the jamb or in the nailing fin, near the top of the frame.

On most units, screwing the sill down is easy because the shims are already in place, but on units over 10 ft. long I use a string to check that the sill is perfectly straight *(Figure 4-13)*. Fluorescent nylon string is common on construction sites, but it's too thick for fine work—just as a framer's pencil is too thick for finish carpentry. I use 20-lb. cotton fishing line—the backing on my fly reels. Fishing line is thin and strong. After stretching it tightly across the top edge of a long sill, I fasten down the sill slowly, making slight adjustments with additional shims wherever necessary (see "Shooting Control Lines," next page).

Correct Cross-Legged Jambs

After the sill is set, I return to the legs and shim each side jamb until it touches my level evenly. Then I secure it with more half-driven nails or screws *(Figure 4-14)*. If possible, I wait until the doors are in the unit to fasten the legs permanently. Levels aren't always perfect so measuring diagonals helps to square a frame. But fitting the frame to the doors is the easiest and best way to ensure that the jambs legs are set properly.

If the doors can't be installed in the frame (on many homes, to save the doors from the plasterers, we set the frames before the stucco, but we don't hang the doors until after the color coat is applied), I stretch strings diagonally from corner to corner, to check that the frame is in one plane and that the jamb legs aren't "cross-legged" *(Figure 4-15)*. A

Figure 4-14. *Shim the side jambs until they touch the level and are perfectly straight.*

Shooting Control Lines

Thankfully, most sliding doors have adjustable wheels. But some units don't, and these sills must be set with absolute perfection. I use a laser or builder's level to install these sills, so that both ends will be perfectly level. Then I stretch fishing line to be sure the sill is absolutely straight.

I also use a laser or builder's level to shoot control lines whenever I'm setting adjoining units, or a series of doors and windows in a long wall, across a large room, or in a custom home *(see photo)*. With the laser or builder's level set in one location, marks can be shot onto the trimmers beside every opening, both interior and exterior, and these marks make it easy to maintain level and straight jamb heads throughout a home, so that head jambs and head casings are perfectly aligned.

Control lines can be located at any distance from the floor, as long as all the lines are level throughout the job site. Once the lines are shot in and marked on the framing, the highest point of the floor (shortest measurement from the control line to the rough floor) is easy to find. From the highest point of the floor, the distance from the control line to the top of the highest jamb can be determined. Afterward, simply match the distance from the control line to the top of that jamb throughout the job.

To ensure that all sills will be level and all head jambs perfectly aligned with each other, I shoot control lines throughout a house. It doesn't matter how high the lines or marks are from the floor, as long as they're perfectly level. I measure up from the line to the top of the jamb—at every opening.

Figure 4-15. *Correct for cross-leg: A string pulled taut from each corner of the jamb should just touch itself at the center of the opening.*

cross-legged jamb can cause a wide sliding door to rub against the window stop. No amount of wax will solve that problem. A cross-legged jamb will also prevent a pair of swinging doors from meeting and sealing flush at the center of the jamb. And a cross-legged jamb will prohibit a single swinging door from lying flat against the stop on the strike jamb *(Figure 4-16)*.

To straighten a cross-legged jamb, I pull the bottom of one leg of the jamb a little more inside the wall and I push the bottom of the opposite leg a little outside the wall — until the strings just touch each other in the center of the frame. I never adjust for cross-leg at the head because that's not where the problem originates, and moving the head of the jamb will always cause problems later when it's time to join the miters in the casing.

Plumbing Jamb Legs

Years ago, I installed an Andersen unit with a pair of 8-ft. swinging French doors. The wall was almost 3/4 in. out of plumb in 8 ft., but I set the jamb flush with the wall, to eliminate the need for tapered

Figure 4-16. *Correct cross-legged jambs at the bottom of the wall—since that's where the problem is—not at the top of the jamb.*

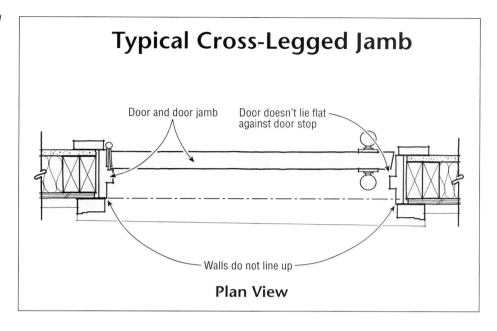

jamb extensions and to make it easier to apply interior casings. Unfortunately, the doors wouldn't stay open at 90 degrees to the opening: They either swung back to the wall or closed by themselves. Since then, I've always checked that the frame is plumb in both directions *(Figure 4-17)*.

INSTALLING SLIDING DOORS

Whether doors slide or swing, once the frame is tacked into the opening, I like to install the doors (if possible) before permanently fastening the jamb legs. On sliding units, the stationary door almost always goes in first. On metal and vinyl sliders the stationary door has to be lifted, with the top of the door going into the milled pocket near the center of the jamb *(Figure 4-18)*. The bottom is placed over the track in the sill, and then the stationary door is slid back against the jamb. To protect the sill from scratches, lift the weight of the door off the track while sliding it toward the jamb.

In some cases, an additional threshold is snapped into the sill, between the front of the stationary door and the strike jamb. This threshold locks the stationary door in place. If there isn't a locking threshold, then screws or other hardware are included with the unit to secure the stationary panel. The active panel is installed next, top first, and then you set the wheels over the track.

On a wood slider, a narrow length of stop — screwed to the head jamb — must be removed before the doors can be installed. However, be sure to replace the stop after installing the doors, or

Figure 4-17. *Check the wall for plumb, too. Swinging doors set in out-of-plumb frames may open or close by themselves.*

they'll fall right out of the frame *(Figure 4-19)*. Wooden sliders are usually placed on the sill initially, and then the doors are tilted up into the opening. The stationary door is installed first, and it slides into a pocket against the stationary strike jamb. The active door goes in last. Be sure the wheels engage the track correctly before tilting the door up into the frame. Then secure the active door with the doorstop. Before sliding the door, adjust the wheels so that the door doesn't scratch the sill.

I wait to fasten the side jambs permanently until I've adjusted the wheels on the active door, so that I can match the jamb to the door. Some wheels have adjustment screws that have to be reached through the leading edge or back edge of

Figure 4-18. *Lift a vinyl sliding door into the head-jamb track first, then up, over, and down into the sill track.*

Figure 4-20. *Install additional shims behind the lock strike and bring the jamb evenly against the door.*

Figure 4-19. *A piece of removable stop secures most wooden sliding doors in their frames.*

Figure 4-21. *Stretch a string across the head jamb, shim the jamb to the string, then screw or nail the jamb to the header.*

the door. Other wheels have adjustment screws accessible through the exterior face of the door, just above each wheel. It's a lot easier to turn the adjustment screws if the weight of the door is removed from the wheels. I use a small pry bar to lift the door slightly, and then adjust the wheels until the door is parallel with the jamb. Then I check that the door slides smoothly. While I'm adjusting the door to the jamb I also look at the alignment of the doors at the center locking stile.

The stiles of the active and stationary doors should be even and parallel, and the muntin bars must line up horizontally. If the sill is straight and level, everything should be perfect.

When I'm sure everything is just right, I fasten the side jambs, placing additional shims behind the lock strike as in *Figure 4-20*. (See "Installing Sliding Door Locks and Hardware.") To get the head perfectly straight I stretch a string tightly across the opening, flush with the bottom edge of the head jamb, and then shim the head to the string and secure it with screws, nails, or both as in *Figure 4-21*.

INSTALLING SWINGING DOORS

As I said earlier, I follow the same sequence when I set jambs for pre-fit swinging doors as I do for sliding doors (I also use the same sequence when I install frames without pre-fit doors). Hanging the

Installing Sliding Door Locks and Hardware

Most manufacturers pre-bore for all operable hardware, though often the strikes are not installed. This is one area where reading the instructions and mocking up the hardware really helps. The strikes usually have to be located after the doors and locks are installed. On some locks, especially those on screen doors, I extend the latch in the locked position and slide the door up to the jamb. I mark a small scribe line beneath the open jaw of the latch, which locates the proper position for the opposing jaw on the strike *(Figure A)*.

Other locks are internal, and the strike seats inside the latch. I seat the strike in the latch and slide the door up against the jamb. A sharp point on the back side of the strike will mark the jamb and locate the proper position for the strike. Some strikes have to be mortised into the jamb in order to close the gap between the door and the jamb. I always install the strike on the surface of the jamb first, in order to determine the depth of the mortise. I never mortise a strike too deep, or attempt to close the margin entirely. Adequate room must be allowed for the weatherstripping; too much pressure will jamb the lock.

Figure A. To install sliding door lock strikes, mark the position of the jaw on the jamb before installing the strike.

Figure 4-22. Hang the door in the frame and fit the jamb to the door. Be sure the jamb extends inside the wall far enough for the drywall.

Figure 4-23. Secure the frame with a shim on each side near the head jamb, and then insert a shim behind the top hinge.

doors and adjusting the jamb legs are the only differences. But if the sill is set straight and level, and the jambs are plumb, everything should be perfect. Once the jamb is set in the opening, I hang the door and adjust the jamb to fit the door.

Begin with Shims

The first step is to secure the jamb, and usually a few finish nails won't do the job because a finish nail can't handle the weight of most exterior doors. Besides, for swinging doors, you must stand on the hinge side with the door closed, in order to adjust the jamb and the door to fit properly. It's difficult to nail through the face of the jamb with the door closed, so I start by using shims *(Figure 4-22)*.

All it takes to secure the frame against the weight of the door is one shim on either side of the jamb, right near the top, but a shim behind the top hinge is a must, too *(Figure 4-23)*.

Most problems with pre-fit doors are caused by the weight of the doors pulling on the top hinges and the upper halves of the jamb legs. Manufacturers usually supply pre-fit swinging doors with enough long hinge-screws so that at least one can be installed in each hinge on the jamb.

Figure 4-24. *Drive a long screw through the top hinge and into the trimmer, but don't tighten the screw too much, or the jamb might deflect and ruin the fit of the door.*

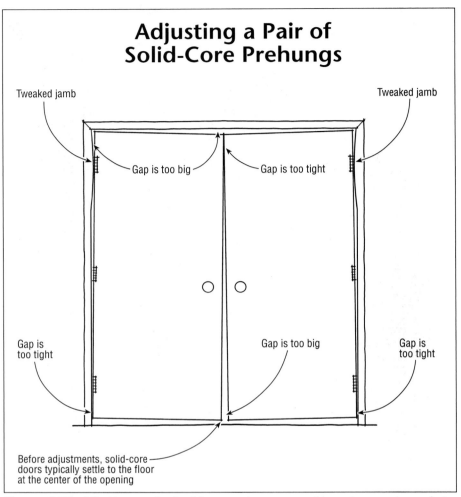

Figure 4-25. *Adjusting a pair of solid-core prehungs seems to require an octopus, but two well-placed screws do the work of four hands.*

Securing Top Hinges

Once I've installed a shim behind the top hinge, sometimes even before hanging the doors in the opening, I drive one long hinge-screw through the top hinge. I don't tighten the long screw too much, or the jamb will deflect. I apply only enough pressure to seat the screw in the hinge *(Figure 4-24)*.

A shim beneath the bottom hinge is also necessary and prevents the lower hinge and jamb from settling against the trimmer. I use the fit of the door to check that the jamb is straight. But I always postpone securing the lower hinge with a long screw until all adjustments are complete.

Adjusting Double-Door Frames

With a long screw securing the jamb, the door should close easily and the gaps between the door and the jamb will probably be correct. However, a pair of doors usually requires a little more adjustment.

Most often, the doors are tight or touching near the top of the lock stiles. In that case, one of the long hinge-screws probably needs to be tightened. Choose which one by looking at the tops of the doors. The doors probably won't be flush at the top, and one will be higher than the other *(Figure 4-25)*. The telling point and solution to the problem is almost always found at the gaps between the hinge-stiles and the jamb, especially above the top hinges. Usually the door that is low will have a much wider top-hinge gap.

Start adjusting the doors by tightening the long hinge-screw in the top hinge on the low door. That adjustment will draw the low door away from high door; it will raise the low door nearer to flush with the high door; and it will decrease the hinge gap

Figure 4-26. *Shim the jamb behind the lock strike, and then above and below the strike, as well.*

Figure 4-27. *Install flashing on the legs first, high enough to meet the top of the head flashing and low enough to cover the drain pan.*

between the low door and the jamb.

If the doors still aren't flush at the top, loosen the long hinge-screw behind the high door until the doors are flush and the gap between the door stiles and the doorjamb is about the thickness of a nickel.

In some cases, a little more back-and-forth adjustment of the long hinge-screws in the top hinges may be necessary to bring the doors into perfect alignment. And occasionally the shims behind the lower hinges will have to be adjusted as well. But all of these adjustments are easy to make if the jamb legs haven't been permanently fastened to the trimmers.

Once the doors are adjusted, install shims and long screws through every hinge. The long screws in the lower hinges can be used to increase the strike-jamb gap, too, particularly near the middle and bottom of the doors.

After the jamb has been adjusted to fit the door, finish shimming the strike side, especially behind the lock strike *(Figure 4-26)*. Place a shim near the bottom of the jamb, too, and then drive nails through the jamb and the shims into the trimmers.

FLASHING

After all the long screws are installed, the jamb can be fastened permanently with nails or screws. But remember, jambs can leak, if not when they're new, then as soon as they begin to settle, so always use flashing.

Flash the Legs First

If the jamb doesn't have a nail-fin or exterior molding, then the flashing can be applied after the jamb is set (see Chapter 5 for flashing a wall before setting a jamb). Install the flashing on the legs first. Be certain that the top of the flashing rises above the frame high enough so that it will reach the top of the head flashing (the head flashing is installed last). At the bottom of the frame, the leg flashing should overlap the sill pan *(Figure 4-27)*. On second-story

Figure 4-28. *Install the head flashing last, in two beads of caulk — one at the top of the flashing and one just above the head of the jamb.*

Figure 4-29. *Even on frames with nail-fins, install the head flashing last. It should overlap the edge of the head jamb.*

doorways, don't fasten the bottom of the flashing or the bottom of the sill pan to the wall because the housewrap must be slid in under those flashings to maintain a proper drain plane where water and moisture will shed down and away from the wall.

Flash the Head Last

Because the housewrap overlaps the head flashing, the head flashing won't have to be lifted later. So set the head flashing above the doorjamb in two beads of caulk, one at the top of the jamb and one at the top of the flashing *(Figure 4-28)*. Even on frames with nail-fins, the head flashing must be installed last, so apply a layer of caulk over sliding doors, too, and then install the flashing *(Figure 4-29)*.

In openings where water penetration poses a serious threat, such as mountaintop homes, beach homes, and such, a second layer of flashing can be applied on top of the nail-fin. Fortifiber markets a self-adhesive flashing for just that purpose: E-Z Seal® (see Appendix 1). The sandwich of Moistop beneath the nail-fin and E-Z Seal® on top of the nail-fin transforms the nailing-fin into a true wide flashing. If there's any concern that siding nails will penetrate and defeat the flashing, a self-adhesive bituminous membrane should be installed on top of the nail-fin.

In addition, on homes without overhangs, an additional drip-edge strip of metal flashing should be applied, to help divert water away from the head of the doorjamb.

HANGING A NEW DOOR IN AN OLD JAMB

There are many ways to hang doors from scratch, but most are time-consuming and frustrating; few guarantee a perfect fit the first time the door is hinged to the jamb. Through trial and error — though mostly through the help of a couple of expert door hangers from a family that specializes in hanging doors — I've acquired a set of door-hanging techniques that are foolproof and fast.

Whether I'm hanging a new door in an old exterior jamb or in an old interior jamb, the techniques are exactly the same. The only difference is whether I'm cutting the bottom of the door for a threshold or for carpet. The examples pictured in this section rely on an interior jamb, which allowed me to keep my head and camera dry while photographing the sequence.

Marking the Door

Since an old existing jamb is almost always slightly out of square, bowed, out of plumb, or out of level, the first step is to scribe the door to fit the opening. Whether the doorjamb is rectangular or arched, perfectly straight or twisted like a pretzel, careful scribing almost always results in a perfect fit the first time the door swings on its hinges.

Position the door for scribing. The first step to scribing a raised-panel door is to cut any excess height off the bottom so that the door can be held

Figure 4-30. *Measure from the jamb to the edge of the stiles to be sure the door is centered in the opening.*

Figure 4-31. *Spread a pair of scribes for the thickness of the floor covering, and scribe the bottom of the door.*

Figure 4-32. *Spread the scribes 3/16 in. to scribe the hinge and lock stile.*

against the opening without projecting above the jamb. This ensures that only a small amount of wood will have to be removed from the top of a door, keeping the width of the stiles and top rail about the same size.

Next, stand inside the stop-side of the opening, pull the door against the jamb, and secure it with a door hook (see "Homemade Door Hook," page 60). Now adjust the door until it's perfectly centered in the opening *(Figure 4-30)*. Start at the head of the door by measuring from the head jamb to the bottom of the top rail at both sides of the door. The dimensions should be the same unless the jamb is terribly out of level. I use a small pry bar to move the door incrementally left or right, and long tapered shims beneath the door to hold it at the right height.

Try to center the lock and hinge stiles, too, so that an equal amount of material will be cut off each side of the door. Again, to ensure that the door is parallel to the jamb, measure from the jamb to the inside edge of the stiles at both the top and bottom of the door. If the jambs are grossly undersized and a lot of wood has to be planed off the stiles, then make sure that the lock stile will be at least 3 7/8 in. wide. This is the minimum width necessary for some locksets (Schlage deadbolts are 3 5/8 in. deep.)

Scribe all four sides. Before scribing the door, I always mark a large X — in pencil or with two pieces of tape — on the hinge stile near the top of the door. This prevents dumb mistakes, like hinging the door backwards or upside down, or drilling for the lockset 36 inches from the top of the door instead the bottom.

For scribing, I prefer to use a set of $2 scribes made by General Tool Co. Although any pencil fits them, I like to use a mechanical lead pencil, because a broken lead is easy to fix while standing on a ladder with one hand braced against the wall and a flashlight in your mouth.

Adjustable scribes are also handy at the bottom of the door, where they can be spread to accommodate the thickness of the floor covering, which varies from 1/4 inch for vinyl to 1 3/8 inches for carpet *(Figure 4-31)*. For exterior doors, you'll also need to know what type of door shoe and threshold will be used.

I like to leave a gap around the door slightly smaller than 1/8 inch. To scribe the stiles, I spread my scribes 3/16 in. apart because I bevel the lock stile and the hinge stile so that they won't rub or bind on the jamb (making the door "jamb bound"), and the leaves of the hinges will never touch (making the door "hinge bound"). The 3/16 in. spread of my scribes works well for a typical 3-degree bevel, which grows almost 1/8 inch longer on the hinge side (long point of the bevel) than on the stop side (short point of bevel), where the scribe marks are made *(Figure 4-32)*.

Be careful to hold the scribes perpendicular to the jamb and press just hard enough to leave a clean sharp line — pressing too hard might accidentally close the scribes. If the grain in one area interferes with the lead, scribe in the opposite direction.

At the head jamb, squeeze the scribes com-

Homemade Door Hook

To scribe a door, the door has to be held tightly against the jamb so it can't move at all, not even a little. It's virtually impossible to do this alone, so I use a homemade door hook made from aluminum channels connected by a length of rubber inner tube *(Figure A)*. One aluminum channel slips over the top of the door, the other catches on the head jamb, and the rubber stretches across the head to hold them in tension *(Figure B)*. One door hook at the top of the door and two shims at the bottom will hold any door snugly against the jambs.

The aluminum hook that goes around the door should be 1 3/4 in. deep so that it will fit both standard door thicknesses. (For thicker doors, I use a door hook who is paid by the hour and helps me carry the door, too.) To help the jamb hook grab securely, I use a grinder to cut small serrations in the edge, and then file and sand the cut edges to prevent scratches in the jamb. The piece of bicycle inner tube should be about 8 in. long.

With the rubber attached, the distance between the two hooks shouldn't be more than 3 1/2 in. for standard interior jambs that are less than 5 in. wide. The rubber is threaded through a slot in the jamb hook: This avoids fasteners that could scratch the jamb. At the door hook on the door, however, I use 3/16x1/2-in. hex-head bolts and a strip of aluminum to clamp the rubber. The excess inner tubing sticks out the top of the door hook. When I work in a house with 6-in. jambs, I loosen the hex-head bolts and slip the excess rubber out a little, which allows for wider jambs. Finally, I mount a knob to the jamb hook. The knob makes it easier to stretch the inner tubing across the jamb.

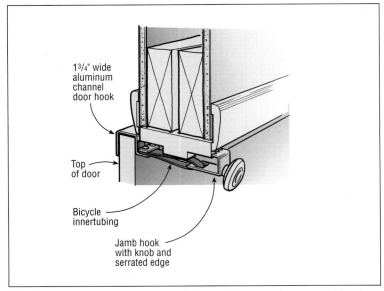

Figure A. *A door hook is essential for scribing new doors to old jambs.*

Figure B. *Pull the door against the jamb and secure it with a door hook.*

pletely closed, because the top of the door doesn't get a bevel. If the top of the door doesn't reach the head of the jamb, spread the scribes so that they just reach the top of the door where it's farthest from the jamb. This ensures that you will remove a minimal amount of material from the top rail.

Locate the hinges and locksets. To avoid confusion and save time, mark the hinge and lockset layout while the door is standing near the opening. To locate the hinges, pull a tape measure down the hinge side of the jamb *(Figure 4-33)*. Press the top of the tape against the top of the jamb, then slide it down almost 1/8 inch—the standard gap between the top of the door and the top of the jamb. You can measure that distance as the tape passes the first hinge mortise—a typical top-hinge mortise will measure 7 in. from the top of the jamb, but only 6 7/8 in. strong from the top of the door. After sliding the tape measure down, hold it tightly with one finger, just below the existing top hinge. The measurement to the top of each existing hinge mortise is exactly where they should be placed on the new door. Write those measurements lightly on the face of the door near each hinge location, or on a scrap of paper.

To measure for the lockset, measure down the strike side of the jamb to the center of the strike and subtract 1/8 inch for the head gap. Again, note

Figure 4-33. *Pretend your tape measure is the door, hold it down from the head jamb 1/8 in., and measure to the top of each hinge.*

Figure 4-34. *To prevent tearing out the grain on the top outside edge of the stile, plane in from the far end of the door with the plane upside down.*

Figure 4-35. *Finish planing the top with the tool right-side up, zeroing-out the cutter as you approach the far-edge of the door.*

the dimension on the door, near the location of the lockset.

On raised panel doors, I like to center the lockset in the lock rail. If the old strike mortise in the jamb doesn't align with the lock rail of the new door, I'd rather move the existing strike mortise on the jamb than have the lockset look out of center in the door. On a painted jamb, I'll fill the strike mortise. But on a stain-grade jamb, either the door must be drilled to match the strike mortise in the jamb or the jamb leg must be replaced!

Cutting, Drilling, and Planing

I do all cutting, drilling, and planing of the door on a workbench especially made for the purpose (see "Portable Door Bench," page 63). The bench holds the door either flat or on edge, and it also provides a storage area for my door-hanging tools.

Because circular saws cause tear-out and chip end grain, use a square and a utility knife to score a line across the face of the lock and hinge stiles. For flush doors with veneered skins, score a line completely across the door. On stile-and-rail doors, I prefer to use a square rather than a straightedge because most head jambs, whether old or new, often have a belly or bow that a straightedge can't follow. On flush doors with veneered skins, a straightedge is the best choice. I also score the far edge of the door, where the saw blade will exit, to eliminate tear-out on the back of the stile.

The top must be perfectly straight. While anyone accustomed to a circular saw can cut the bottom of the door quickly, cutting the top isn't so easy. The top of the door has to be cut perfectly straight—if the saw tips even slightly, the gap between the top of the door and the jamb will be uneven. One trick is to cut just outside the pencil line, then use an electric door plane to finish the cut right to the line. To eliminate tear-out at the end, start by holding the plane upside down and come in from the far side of the door *(Figure 4-34)*. It's essential to keep pressure on the fence beneath the door to make sure the plane stays square with the top of the door. But because this is awkward, I plane only about 6 in. into the door right to the pencil line. Then I turn the plane around, right-side up, and come in from the opposite end of the door to finish the cut *(Figure 4-35)*.

The hinge stile gets a slight bevel. To avoid costly mistakes, the "X" and all the scribe lines must always point toward the inside of the door bench. To keep this straight, imagine that the

EXTERIOR DOORS 61

Figure 4-36. *Follow the scribe line and adjust the depth of cut with your thumb while planing the door stile.*

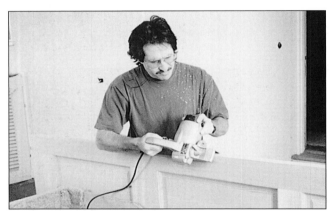

Figure 4-37. *Turn the plane at an angle to ease the edges of the door stile.*

Figure 4-38. *The bevel must be consistent and fall away from the square a little more than 1/16 in.*

passes that slowly approach the scribe line. I like to just leave the line, and then finish by tipping the plane at an angle and easing the edge of the stile *(Figure 4-37)*.

The lock-stile bevel is essential. To avoid having to move the door too many times, I mortise the hinges before planing the lock stile. But the lock-stile planing process is the same. The biggest hurdle is cutting a consistent bevel. Otherwise, the gaps will be too big, or the door will rub against the jamb or may not even shut. Check the angle frequently by placing a square across the bevel at various points along the stile. On a 1 3/4-in. door with a bevel of about 3 degrees, the bevel should fall away from the square a little more than 1/16 in. *(Figure 4-38)*.

Mortising and Boring the Door

When I was young I used to watch my father's finish carpenter hang doors. He used hand planes, chisels, and steel drill bits, which he turned with a brace. Every morning, after pouring steaming coffee from his thermos, he patiently sharpened those chisels and bits. I haven't met a contractor yet who's willing to pay me to chisel mortises by hand, let alone sharpen my chisels on his time. These days I use machine tools for all mortising and boring operations, and door hanging moves at a faster pace. But a few simple rules reduce the risk of costly errors.

Mortising for the hinges. The first rule when laying out hinges is to look for the X, to avoid hinging the door upside down. Find the X and the top of the door, pull a tape measure down the edge of the door from the top to the bottom, and then transfer the measurements taken earlier from the jamb onto the door *(Figure 4-39)*. Use a sharp pencil and a square to extend the marks straight across the edge of the door.

bench is just like the doorstop on the jamb. The hinges will then be installed with the barrels pointing away from the door bench. Also, the door plane should always bevel in the same direction: down toward the inside of the bench.

The Porter Cable 126 door plane I use is meant exclusively for doors. The depth-of-cut adjustment lever—which is the best I've ever found—makes it easy for me to hold the handle and trigger in my right hand, and use my left thumb to adjust the cutter depth *(Figure 4-36)*.

Watch the scribe line carefully. If it isn't parallel to the edge of the door, slowly plane until the line and the edge of the door are parallel, and then plane closer toward the line with successive passes. Again, it's important to hold the fence tightly against the face of the door, so that the plane doesn't rock. Otherwise, the bevel might vary and cause an ugly wave in the margin between the door and the jamb. Never bury the cutter or try to cut too much on one pass. There's no hurry, so just make smooth

Portable Door Bench

Many manufacturers make a simple clamping device to hold a door securely while you work on it. But I hang a lot of doors and I'm too old to bend over all day picking up tools, so I use a door bench that not only holds a door either flat or on edge but carries all my tools as well. My 22x68-in. bench is narrow enough to fit through a 28-in.-wide doorway, but long enough to support an 8-ft. door. It's designed so that I can step inside it to carry it around.

I'm 5 ft. 7 in. tall, so my bench is 32 in. high, a comfortable working height that is still short enough so that when I step inside the box to carry it the legs clear the ground easily. Also, at this height I can climb stairs while walking inside the bench without having to lift it up over my head.

The bench has adjustable rungs to hold doors of varying widths on edge. The best layout I've found is to have four rungs spaced 5 in. apart. The 2x4 legs can be drilled to accept 1 1/4-in. closet pole, but doors slip off round rungs too easily. To avoid having to put stops on the ends of round rungs, I made rectangular rungs that fit into slots in the laminated hardwood legs of my bench *(Figure A)*. Since I store my bench in the back of my van, I attached the legs with butt hinges that fold flat. When I carry the bench up to my van, I set the front end in the back of the van and push the bench in until the front legs fold up, almost automatically *(Figure B)*. Then I step out of the box, fold the back legs up, and slide the bench in the rest of the way. I carry a 1x2 spreader to secure the legs open while I'm working at the bench.

Flanking the central opening are two 22-in. tool shelves with dividers to hold the tools securely. The sides of the bench are deep enough to hold the tallest tool while still allowing a door to lie flat across the top of the box and not rock on a tool handle.

The door bench also has an adjustable steel hook, which clamps doors on edge. The hook moves in a mortise, like a bolt in a track, and is locked in position by means of a single wing nut *(Figure C)*.

Figure A.
Four rectangular rungs provide a stable surface for holding doors, from 36-in.-wide doors on the bottom rung to 12-in.-wide sidelites on the top rung.

Figure B.
Folding legs: Mounted to the bench on hinges, these legs fold neatly for storage in a truck or van.

Figure C.
The hook: Mortised into the front face of the bench, this hook is adjustable for doors of nearly any thickness.

Second rule: Always draw a light X beneath each measurement, so that there's no confusion about where to set the hinge template. This way, the hinges will never end up on the wrong side of the mark *(Figure 4-40)*.

I hang a lot of doors, so I use a router and templates to cut the mortises *(Figure 4-41)*. The templates are made by Templaco Tool Co. (see Appendix 1) and include almost every hinge size and almost every common type of hardware, including flush bolts and mortise locks.

First, set the router to the right depth by holding the hinge flat against the bottom of the template and extending the bit until it's flush with the face of the hinge. Then attach the template to the door, positioned above the first layout line by 1/16 in., which is the clearance of the router bit guide. Rout just a small nick of the first hinge mortise, and then check the depth setting before going further.

To avoid tearing out the face of the door, first run the router into the door along each shoulder of the template. Then make a pass along the face

Figure 4-39. *Look for the X, and then measure down from the top of the door for each hinge location.*

Figure 4-41. *Use a template and router to mortise for hinges.*

Figure 4-40. *To prevent cutting hinge mortises on the wrong side of the line, mark an X on the right side of the line.*

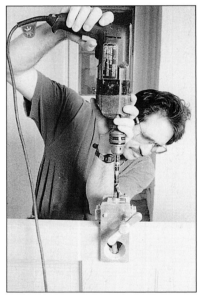

Figure 4-42. *A lock-boring jig is like a portable drill press.*

of the mortise, and finish by cleaning out the center. For a smoother cut, always move the router against the rotation of the bit. For square-cornered hinges, after routing, I use a corner chisel to square-up each mortise.

I always drill pilot holes for hinge screws. Four good-sized screws driven into a 1x3-inch space act like a wedge and, without pilot holes, will split the door and ruin the job. I use a Vix Bit® (see Appendix 1) to speed up the drilling.

Boring for the lock. Once again, look for the X first, and then measure down from the top of the door and mark the lock location. Always drill the face-bore first, and then the edge bore. This takes the guesswork out of the edge bore—when the bit falls into the hole, you're done.

A hole saw and a spade bit will do the job, but I prefer to use a lock boring jig, such as those made by Porter Cable, Classic Engineering, and Templaco Tool Co. (see Appendix 1). A boring jig is fast and accurate *(Figure 4-42)*, and will also ensure the proper backset—either $2^{3}/_{8}$ in. or $2^{3}/_{4}$ in., depending on the hardware. If you don't have a boring jig, layout the face bore using a square or the paper template that came with the lockset.

Mortising for the latch. I use a router and templates for all latch and strike mortising, too. The router is faster than a chisel, and the template ensures that each mortise will be the perfect size and depth. Most residential latches measure 1x$2^{1}/_{4}$ in., so only one template is necessary. (Schlage makes a deadbolt with a wider $1^{1}/_{8}$-in. latch face, though it's rarely used in residential work). Although Templaco manufactures templates along with plastic locators, I sometimes make my own templates and use a line scored across the center to align the template over the edge-bore. Adjustable stops on the bottom of the template keep it centered over the edge of the door.

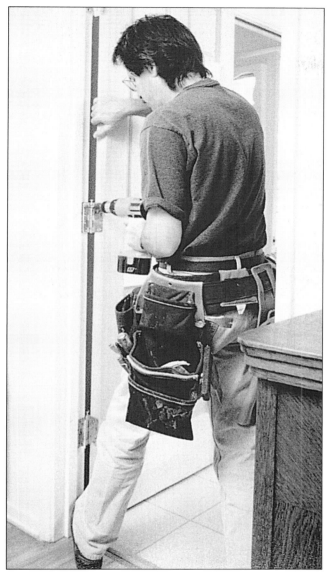

Figure 4-43. Hang the door up on the top hinge and hold it with one foot while attaching the lower hinges.

Before turning on the router, place it on the template with the bit inside the edge-bore hole. Cut the outside of the mortise first, and then clean out the center. To avoid nicking the template, lift the router straight off the template without tilting. Use a corner chisel to square up the rounded corners left by the router.

Adjusting the Hinges

To hang the door, first split the top hinge. If the existing hinge-screw holes don't line up with the new hinges, and use the top hinge leaf to drill pilot holes for all three hinges. Then attach the leaf to the top mortise on the jamb. Lift the door onto the top leaf and drive in the hinge pin, and then swing the door open and hold it still with one foot while driving a hinge screw into the middle and bottom hinges. Attaching the lower hinges to the jamb while they're on the door ensures that they'll be aligned perfectly *(Figure 4-43)*. If a door is scribed carefully and beveled correctly, it will usually fit the first time up. If it doesn't fit just right, small hinge adjustments will probably correct the problem. I always adjust the hinges before installing the lock strike, because the position of the strike might change.

The secret to adjusting hinges is to imagine the door and the jamb as a rectangle within a rectangle. If the rectangles aren't parallel, one of them — the door — has to be rotated *(Figure 4-44)*.

Spreading hinges. When the head gap is too tight or the top of the door rubs the head jamb, the gap at the top of the lock stile is usually too big. The quickest remedy is to bend and spread the top hinge, which rotates the door down.

To bend and spread the top hinge, open the door enough to slip the head of a nail set between the leaves of the hinge, and then pull the door closed and gently spread the hinge *(Figure 4-45)*. Don't pull too hard or the hinge screws might rip out. Bend the hinge a little at a time and check the fit each time.

If the head gap is too big at the lock side, the gap at the bottom of the lock stile will generally be too big, as well. In this case, bend and spread the bottom hinge to rotate the door upward.

Squeezing hinges. In the reverse situation, the lock stile rubs on the jamb. This means the hinge gap is too big, so one or more hinges may have to be squeezed shut a little, which will pull the door back towards the hinge jamb.

Use an adjustable crescent wrench to bend and squeeze the top hinge. Drive the hinge pin up until only the top knuckle is engaged, and then tighten the wrench on the top knuckle of the door leaf—not the leaf attached to the jamb *(Figure 4-46)*. Bend the top knuckle on the door away from the jamb, toward the lock stile of the door, then carefully bend the other knuckles on the door leaf the same amount. This adjustment will pull the door tighter to the hinge jamb, increasing the gap on the lock side.

Bright brass hinges dent and mar easily. To adjust one of these, instead of spreading or squeezing, remove it from the jamb and either shim behind the hinge slightly or chisel the mortise a little deeper.

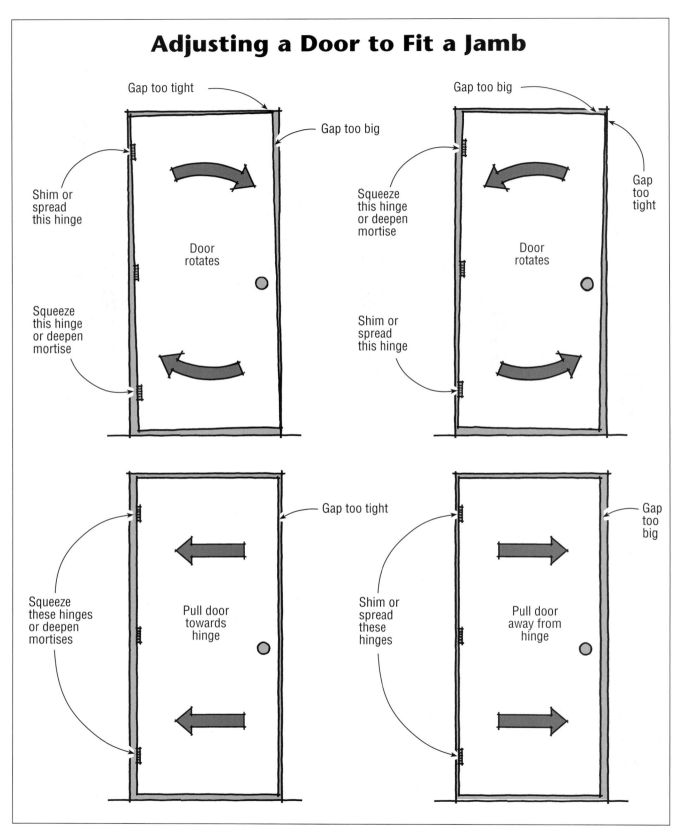

Figure 4-44. Even if careful scribing doesn't always result in a perfect fit, the door can be "rotated" in the opening by spreading or squeezing the hinges (top left and right). If the gap against the lock jamb is too tight or too big, squeeze or spread the top and bottom hinges to move the door toward or away from the hinge jamb (bottom left and right).

Figure 4-45. *When a door's head gap is too tight, slip a nail set between the leaves of the top hinge and gently pull the door closed to slightly spread the hinge.*

Figure 4-46. *If the door is too tight at the lock stile, use a crescent wrench to squeeze the top hinge by carefully bending the knuckles on the door leaf of the hinge. Bend the knuckles away from the jamb, toward the lock stile of the door.*

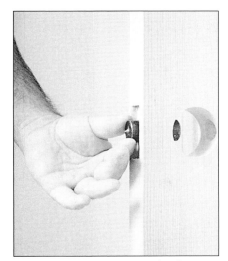

Figure 4-47. *A center marker locates the exact position of the lock strike on the jamb — almost.*

Figure 4-48. *Place the strike over the latch and mark the edge of the door.*

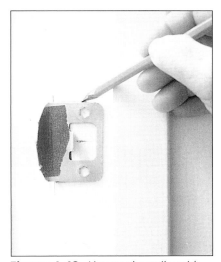

Figure 4-49. *Line up the strike with the vertical and horizontal pencil marks and trace the outline of the strike onto the jamb.*

Locating the Strike Plate

Most boring jigs come with a center marker, which is the quickest way to position a lock strike. The marker is a simple steel cylinder that fits in the latch bore. The sharp point centered on the face makes a dimple on the jamb, which marks the spot to drill for the latch *(Figure 4-47)*. But this method is not always exact, and I use it only if the strike plate is "adjustable" (has a thin piece of metal projecting at right angles from the back of the latch hole). If the latch fits loosely in the strike, this tongue can be bent slightly to snug the door up against the stops.

While I use a center marker most of the time, for

EXTERIOR DOORS 67

stain-grade jambs I prefer to install the lockset and use the latch to find the exact location for the strike. Stick a small piece of masking tape on the strike and place the strike over the latch, which should just touch the shoulder of the strike *(Figure 4-48)*. Use a pencil to mark a line down the face of the strike where it meets the face of the door. (The masking tape will allow that line to show up more distinctly than if it were drawn right on the shiny surface of the finished strike.) Next, shut the door and mark a line on the jamb with the pencil riding the face of the door, and then mark the jamb at the horizontal center of the latch. Open the door and extend the horizontal mark across the face of the jamb, then hold the strike centered on that horizontal line. Finally, line up the vertical pencil marks on the strike and the jamb, and trace the outline of the strike on the jamb *(Figure 4-49)*.

Again, I prefer to use a router and template for the strike mortise because that's cleaner and faster, and it eliminates having to hammer on the jamb, which could change the margin between the jamb and the door. If I've done everything right, the dull "thunk" of the latch falling into the strike just as the door comes up against the stop is sweet reward.

CHAPTER 5

Windows

Ordering windows has become a specialized task. A multiplicity of manufacturers offers an endless variety of window styles in different compositions and sizes. Code requirements for windows, from minimum egress sizes to locations for tempered glass, are also complex and change frequently. It's no surprise that ordering the right windows has become a challenge for many contractors. Having made many errors myself, I've developed a useful form for window takeoffs and orders. My form includes categories that fulfill the requirements of most manufacturers (see the chart "Window Takeoff," Appendix 2).

ORDERING WINDOWS

Ordering windows is similar to ordering prefit door units, but windows generally require more thought and foresight. There's a lot of information that the manufacturer, client, and contractor need to know. Reviewing and completing each of the categories included in the take-off form should help to prevent errors and omissions on your window orders.

Don't be tempted to combine like-size units until you're certain that they are exact duplicates. Glass types can vary significantly: bathroom windows might need to be obscure; windows within 12 in. of a door or 18 in. of the floor might have to be tempered; jamb widths can vary, too, according to shear wall (sheathing) specifications. So wait until you have filled in all the details for each window opening before combining like-size windows within one entry.

Specify the manufacturer and window type. Most plans specify a manufacturer, and if the plan doesn't, I always do. I submit my order forms for final client/contractor approval, so including the manufacturer is a good idea. At the head of the form I also specify the type of window, whether it's wood, vinyl, clad, primed, or stain-grade.

Identify each window and wall thickness. Blueprints can be confusing, especially when architects include every bit of information on the floor plan. To simplify and clarify a busy set of plans, I use colored markers to fill in the window and door labels — one color for windows, another for exterior doors, and a third for interior doors. Beside each exterior wall I note the finished wall thickness in red ink. This number represents the combined thickness of stud size, shear wall (sheathing), siding/stucco, drywall/wetwall, etc.

Make a "walk-through" list. After coloring in the window locations, I turn to the window schedule and make a list of the sizes and types on a legal pad. Then I study the plan carefully — walking through the actual job site, if possible, and if not, then walking through it mentally — and I list each window by its size and type, with a subheading under each window size for different wall conditions, types, glazing, etc.

Beside each window type, I note whether the window is a fixed, double-hung, casement, slider, bay, or arched type, etc. If the window is a casement or slider, I also note the hand — the side where the hinge or active slider is when looking at the window from the outside of the house *(Figure 5-1)*. Finally, because windows and doors often change between the plans and the as-builts, I note which room (bedroom, bathroom, kitchen, etc.) each window belongs in. This way, they're easier to identify and making changes isn't difficult.

Check the glass type. While most windows

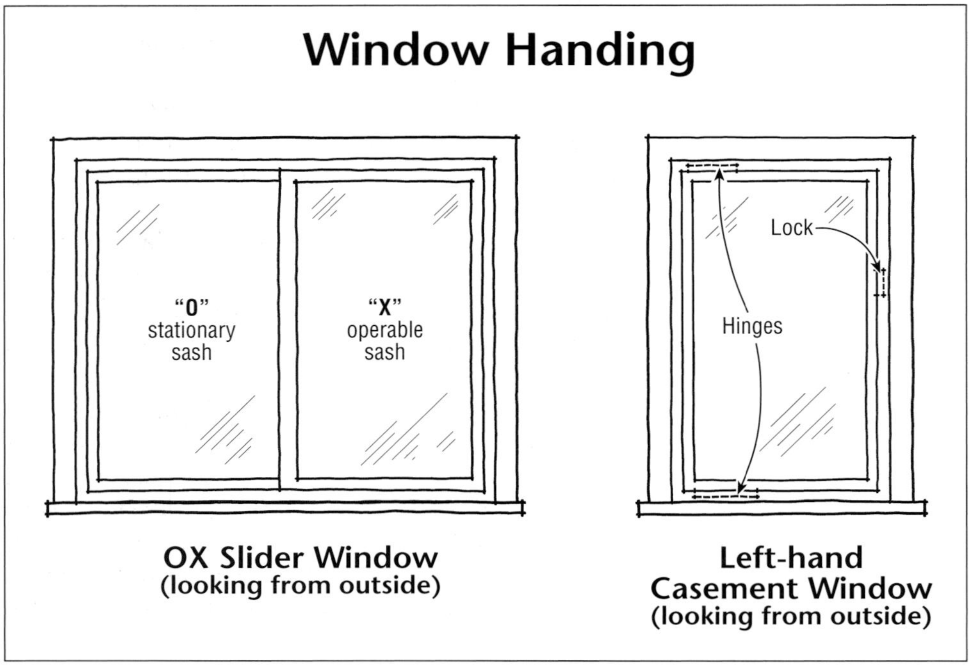

Figure 5-1. *Just like doors, windows are "handed." But for windows you almost always stand outside the house to determine the handing.*

Figure 5-2. *An Arts and Crafts double-hung window typically has a single lite in the lower sash and a pattern of smaller lites in the sash above.*

today are dual glazed, glass type can still vary throughout a house. Tempered glass must be used within 12 in. of a door, in showers, or within 18 in. of the floor. Obscure glass is often the choice for bathrooms, while art glass (leaded, tinted, stained, beveled, sanded, etched, blue chip, etc.) is frequently used in entry doors and sidelites.

Carefully note the lite design. Under the lite category I note whether the windows are single-lites (each with a single, clear, full-size pane of glass) or cut-ups (having individual lites or small panes of glass). If the windows are cut-ups, I note if the muntin bars are applied grids on the surface, if the lites are sandwiched between the panes of glass, or if the cut-ups are true divided lites (TDL). I also note the cut-up layout.

On some projects, identifying the correct cut-up layout can be time-consuming. Homes designed to emulate historic period styles require specific layout designs. Georgian and Federal homes are known for their symmetrical window-pane layout, which is often six-over-six (six panes of glass in the upper sash and six panes of glass in the lower sash of a double-hung window) or nine-over-nine. Today's manufacturers describe this latter window as 3w/3h (3 lites wide by 3 lites high per sash). Arts and Crafts windows often have a single large pane of glass in the lower sash, topped by small panes of glass in the upper sash *(Figure 5-2)*. Modern homes frequently require custom cut-up designs and dimensions so that muntin bars in French doors and flanking windows will align horizontally.

Include the rough-opening size. Of course I include the rough-opening size for each window so that the window takeoff can be supplied to the contractor and the framer, which helps to ensure that the framing is correct. If the framing contractor installs the windows, he'll have no trouble determining the correct window for each opening. If I install the windows and the framing isn't correct (I often need to perform surgery on rough openings), then the takeoff list helps document my extra charges.

INSTALLING WINDOWS

Practices vary across the country, and sometimes framing contractors install windows and exterior pre-fit doors and jambs. But framing is called rough carpentry for good reason. As a finish carpenter, I prefer to set the windows, doors, and exterior jambs myself. This way, I know that window and door casings are parallel and align horizontally throughout a house; that window and door jambs project an even amount inside the home and proud drywall isn't in the way when it comes time to apply the casing; and finally, that door jambs are plumb, straight, and never cross-legged.

Waterproofing Is First

Most construction specialists agree that windows and doors leak. As building consultant and author Joseph Lstiburek says, "Windows and houses are like people — the more they age, the more they settle and the more they leak." Even though many manufactured units are watertight when they're new, most will leak as they age and settle. Leaks should be anticipated, and adequate drainage should be planned in advance of window and door installation (see Chapter 4, "Exterior Doors," page 47).

Start at the bottom. After checking that the rough opening is the correct size — about 1/2 in. larger than the window frame in both width and height — I first apply a length of flashing across the bottom of the opening. Waterproofing always begins at the bottom, in shingle-style, so any moisture that penetrates the siding or stucco can be directed to the surface of the housewrap and then outside.

I apply a 12-in.-wide strip of Moistop®, (or other comparable window flashing) across the bottom of the opening *(Figure 5-3)*. I cut the bottom piece of flashing the length of the opening plus twice the width of the flashing, so that the jamb-leg flashing will overlap the sill flashing. I fasten the flashing with a small stapler, but only along the upper edge. The bottom of the flashing must remain loose, so that building paper or housewrap can be tucked up beneath it.

Sill pans for every opening. Before installing the sill pan, I place a small piece of bituminous membrane in each of the lower corners of the opening *(Figure 5-4)*. Next, I install a sill pan that's also made of bituminous membrane. The pan must hang down far enough to lap over the waterproof flashing along the bottom of the window opening. The pan should also extend inside the wall to the face of the studs. Finally, the legs of the pan should climb 6 in. to 8 in. up the trimmers or jack studs.

I slice the corners of the pan and lay the loose flaps down on the patch, which creates a tightly sealed corner *(Figure 5-5)*. The legs of the flaps rise up the trimmers, and the bottom extends about 8 in. on each side of the window.

Install the legs, but not the head. The jamb-leg flashing goes on next *(Figure 5-6)*, and must be long enough to overlap the sill flashing and underlie the head flashing (the head flashing isn't installed until after the window is fastened in place). I fasten the jamb-leg flashing to the trimmers or sheathing and staple both the front edge and the back edge securely because the housewrap, rather than tucking under the jamb leg flashing, lies on top.

Setting the Window

I run a thick bead of urethane caulk on the back of the nailing fins before setting the window in the rough opening *(Figure 5-7)*. (For more on this, see Chapter 4, "Exterior Doors," page 50.) The bead should be large enough to seal the fin to the flashing, but not so abundant that it will squeeze out and make a mess. Vinyl and aluminum windows are manufactured with fins on all four sides, but to allow moisture to drain out from the pan and escape down the outside of the drainage plane, I do not caulk the bottom fin to the flashing.

Installing windows is usually a two-person job. It takes two just to lift a large window into its opening, but even on a small window, a two-person crew works much more efficiently once the window is in place: one carpenter works inside while the other works outside. If I'm working alone, as on a small unit like the one pictured, I have to go inside the house and out in turns. First, I press the window against the rough opening, and then slide it from side to side until it feels centered in the opening. I also lift the window off the sill about 1/4 in. to 3/8 in., which allows a little extra room for window cranks to rotate after the interior wood sill is installed. In addition, that extra space at the sill allows additional room for shimming, which is especially important if the rough wood sill is bowed, bellied, or out of level. Once the window is

Figure 5-3. *Waterproofing is essential. Begin by flashing across the bottom of the window.*

Figure 5-4. *Place a small patch of bituminous membrane in each of the lower corners of the sill.*

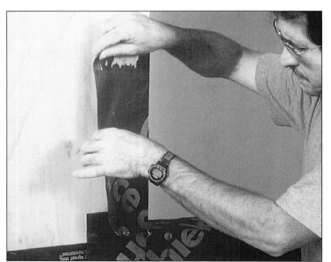

Figure 5-5. *Then lay the sill pan, which should extend 6 in. to 8 in. up the trimmers or jack studs, over the corner patches.*

Figure 5-6. *Next, install the jamb flashing so that it overlaps the sill flashing, and reaches 12 in. above the top of the window opening.*

positioned, I place a level on the exterior molding at the sides and top, and adjust the unit until it's plumb and level. Then I temporarily secure it with a screw or nail at each top corner *(Figure 5-8)*.

Before permanently securing the bottom of the window, I look at the unit from inside the house. I check that the window is centered in the opening and an even distance from the interior face of the stud wall. For wooden windows, I make sure the jamb projects almost 3/4 in. inside the house (I order my window jambs with a 3/4-in. projection for 5/8-in. drywall because 5/8 in. is never enough to keep the drywall from being proud of the jamb). I use a small pry bar to pull the window into the house until it's flush with a 3/4-in. spacer block, then shoot a nail through the face of the jamb into the trimmer or jack stud *(Figure 5-9)*. The pry bar helps overcome the resistance of the flashing paper, or sometimes even that of the sheathing if it's not nailed off tightly to the framing. The spacer block ensures that the window jambs will extend far enough inside the wall, but not too far. As I said earlier, I'm not a framer and I don't just pull the window into the opening until the exterior molding feels like it's hitting the outside of the house. I check that the window is just right before securing it to the framing. Then I go back outside.

Securing the Window

Once the window is centered, plumb, and level, I install one screw in a bottom corner, then take diag-

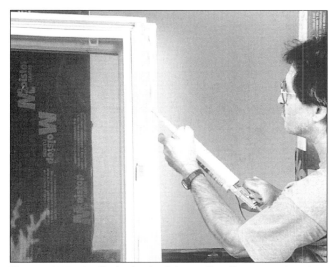

Figure 5-7. *Caulk the back of the nailing fin or exterior trim molding on the sides and top with urethane caulk; leave the bottom fin uncaulked.*

Figure 5-8. *Secure the window temporarily with only one nail (or screw) in each top corner.*

Figure 5-9. *Pull the jamb inside the framed wall so that drywall won't be proud and casing will be easy to install. While holding the jamb with the pry bar, shoot a nail through the jamb and into the trimmer.*

Shimming Vinyl Windows

Because vinyl windows are flexible and climatic conditions can cause slight changes in their dimensions, vinyl window manufacturers often require solid shimming beneath window sill frames. But ripping a continuous shim to fit perfectly into an imperfect space isn't easy. Besides, a perfectly installed continuous shim will impede water drainage from the sill pan.

I carry an assortment of shim stock in long lengths just for this purpose, ripped in odd sizes from $1/8$ in. to $1/2$ in. wide. I snap off pieces of shim material and fill the void between the window sill and the sill pan, but I leave small gaps for water drainage.

onal measurements to be sure the window is square before fastening the opposite corner *(Figure 5-10)*. Most window units are manufactured square, but it's important to check because the sash may not operate properly if the window frame is racked.

On windows under 8 ft. long, I lay a level on the top of the sill to check for bows and bellies. On longer windows, I stretch a tight string across the sill. I use shims to lift the sill up to the level and tack them through the sill *(Figure 5-11)*. If the sill of an aluminum or vinyl window is bowed up, I walk back outside, pull the offending spot down, and install a screw through the nailing fin at that location. Then I check the operation of the window. If the window is a double-hung, like the one in this example, then the gap between the lower sash and the sill must be straight.

If the window is a slider, I remove the screen and slide the operable sash close to the locking jamb. The gap between the leading edge of the operable sash and the face of the jamb must be even, from top to bottom. If the gap isn't even, either the sill isn't straight (a small belly or bow in the sill just beneath the rear wheels of the sliding sash will cause this problem) or the frame of the window is out of square. Usually only minor adjustments are necessary: shims can be placed between the window sill and wood-frame sill, and the two bottom screws can be removed and the frame racked slightly to accommodate the sash. Most sliding windows have adjustable wheels, but the wheels shouldn't be adjusted too much, or the locking stile of the operable sash will not align vertically with the locking stile on the stationary sash.

Figure 5-10. *Check the window for square before permanently fastening the exterior molding to the trimmers.*

Figure 5-12. *Run two beads of caulk above the window and press the head flashing into both (above). Lap the edge of the flashing over the top of the exterior window molding to ensure a complete seal (left).*

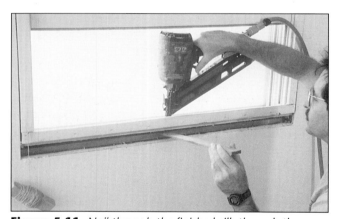

Figure 5-11. *Nail through the finished sill, through the shim, and into the rough sill.*

Unlike vinyl windows (see "Shimming Vinyl Windows," previous page), aluminum and wooden windows have little problem with extensive sagging, expansion, or contraction, and a few shims, placed about every 8 in. beneath the sill, are all that's needed before permanently fastening the nailing fins. I install a screw or nail about every 12 in. around the frame. On windows with wooden sills, I nail through the finished sill, through the shims, and into the rough sill. For windows with nailing fins beneath the sill, I try to install as few fasteners as possible, so that moisture or water that collects in the sill pan can escape.

Installing the Head Flashing

Installing head flashing is the last step in window installation and is one of the most important steps. Many windows are not made with an integral, waterproof top cap, so the head flashing must lap over the top edge of the window frame. Self-adhering membranes work well for this job, though I often use simple Moistop® instead. First, I run a generous bead of urethane caulk on the top of the window frame and nailing fin. Then I run a bead of caulk on the sheathing or header, near where the flashing will end. Last, I firmly press the flashing paper into both beads of caulk and staple it, too, so the wind won't lift it off the sealant *(Figure 5-12)*.

HANGING CASEMENT WINDOW SASH

Whether it's because of tight budgets or a growing interest in saving and reclaiming old building material, I'm often asked — in jobs involving casement windows — to hang old sash in new jambs, or new sash in old jambs. I enjoy this type of recy-

Figure 5-13. *Add a dam to a casement window sill to protect against wind-blown rain.*

Figure 5-14. *Note the position of the hinges on the sash, add $1/8$ in. for the head gap, and transfer the hinge locations to the jamb (top). Use a router and template to cut the hinge gains (above).*

cling, for the opportunity to both save landfill space and work with old materials.

If I need to hang a new piece of sash in an old jamb, I follow the same steps for hanging a new door in an old jamb. First, I scribe the sash to fit the jamb. Then I hinge the new sash to match the hinge layout on the jamb, allowing $1/8$ in. for head gap, just as I do for doors. (See "Planing a Bevel on Narrow Sash," page 77.)

Building a New Casement Jamb

I use a different technique for hanging an old piece of sash in a new jamb. After measuring the old sash and checking that it's square, I build the new frame to fit the sash and add $3/16$ in. to the width and height to allow for hinge, strike, head, and sill gaps. To allow for the addition of a dam on the sill, I add another $1/2$ in. to the height of the new frame *(Figure 5-13)*.

Once the frame is built and tacked together, I measure the hinge layout on the sash, and then transfer those measurements to the hinge-leg of the jamb. To allow for proper head and sill gaps I add just less than $1/8$ in. at each hinge *(Figure 5-14)*.

Weatherstripping Makes A Difference

Weatherstripping windows is slightly different from weatherstripping doors. Several different products can be used to weatherproof a window, and each product affects the fit and size of the window in the jamb.

V-Bronze or spring-bronze weatherstripping was once the most common type of jamb seal. I continue to use V-Bronze today, especially when renovating older homes. I always increase the hinge and head gaps an additional $1/6$ in. to allow for the thickness of V-Bronze *(Figure 5-15)*.

Kerf-in weatherstripping has grown in popularity in recent years, and for good reasons. These vinyl-skinned, thermoset foam-core products work well to keep out rain, wind, and cold, and their elasticity allows for seasonal movement, too. For casement windows, kerf-in weatherstripping is applied to the exterior edge of the window sash. An additional $1/4$ in. must be added to the frame size to allow for this type of weatherstripping *(Figure 5-16)*.

Sealing casement windows. There are several ways to seal the bottom of a casement window. A dam and drip are often installed, along with a variety of other weatherstripping products, but I prefer

Figure 5-15. *V-bronze, once the most common type of jamb seal, is still used, especially for renovation work.*

to treat casement windows just like small doors, and I take care to seal them properly.

I use a 1¼-in.-wide door shoe on the bottom of casement windows. To accept the ½-in. height of the door shoe, I rabbet the bottom of the window on the outside of the sash so that the shoe and drip won't be visible from inside the home *(Figure 5-17)*.

REPLACING DOUBLE-HUNG WINDOWS

Repairing and restoring double-hungs is a job I thoroughly enjoy. When I'm renovating old pulleys, ropes, and lead sash balances, I'm reminded of the history of carpentry and the continuity of the craft. Matching replacement sash, planing these windows smooth, and waxing the frames so that they slide with ease is a fulfilling accomplishment, as well. Besides, after lugging heavy doors, it's a relief to have a small window on my bench.

But restoration and repair is not the by-word of every client. Many homeowners can't afford to revive their old windows and maintain them year after year. Besides, an old, drafty, uninsulated, single-pane, double-hung window feels almost like a new, dual-glazed, insulated double-hung window that's halfway open.

Wooden replacement windows with vinyl jamb liners are popular and easy to install, but vinyl

Figure 5-16. *Kerfed-in weatherstripping is most often applied to the exterior edge of the window sash. Cut the kerf with a slot cutter (top) so the weatherstripping is positioned toward the outside of the sash (middle). When the window is closed, the weatherstripping rubs against the strike jamb (bottom).*

Figure 5-17. *One way to properly seal the bottom of a casement window is to treat it like a small door: cut the shoulder for a door-shoe rabbet with a circular saw (top) and finish the rabbet with a slot cutter (middle). The shoe will not be visible from inside (bottom).*

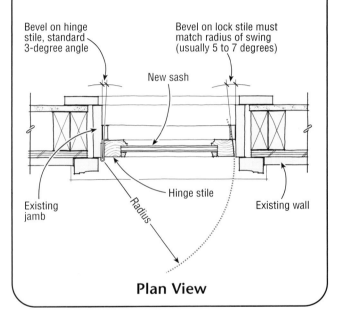

Planing a Bevel on Narrow Sash

I use my Porter Cable 126 planer to put a 3-degree bevel on the lock stile of most doors, but narrow sidelites and sash require a steeper bevel *(see illustration)*. Imagine that the sash is the radius of a circle. The smaller the sash, the tighter the radius. A tighter radius requires a steeper bevel on the lock stile for the short point of the sash to clear the edge of the jamb.

Plan View

replacement windows are far less expensive and represent 90 percent of the replacement work that I do.

Measuring for A Replacement Window

Making a profit in this competitive market is a challenge, but it can be done if you know how to measure for replacement windows correctly. I measure the width of the old jamb right behind the blind stop — at the top of the jamb, the center of the jamb, and the bottom near the sill. I subtract $1/8$ in. from the smallest measurement so that the new frame will fit into the existing jamb *without difficulty* **(Figure 5-18)**.

The exterior of the new frame will abut the blind stop, so the height of the window must be measured against the blind stop, especially where the blind stop meets the sloped sill **(Figure 5-19)**. While it's tempting to subtract more from the size of the window, any gap larger than $1/8$ in. can't be filled with caulk and must be trimmed with wood. Adding more wood to the exterior of a window increases maintenance problems and defeats the purpose of choosing a vinyl replacement unit.

Installing a Replacement Window

The nice thing about replacement windows is that you don't have to finesse the removal of the old

Figure 5-18. To ensure that a replacement window will fit, measure the width across the top, center, and bottom, and subtract 1/8 in. from the shortest measurement.

Figure 5-19. Measure the height right behind the blind stop, to the long point on the sloped sill.

Figure 5-20. If the space is too tight for the replacement window, cut back the stool with a rotary cut-out saw, or a panel saw.

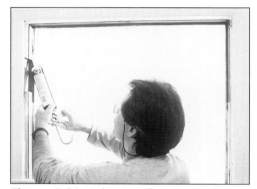

Figure 5-21. Before installing the window, scrape the old jamb clean and caulk the blind stop.

Figure 5-22. Install the window with countersunk screws and cover the heads with adhesive screw cap.

sash. The sash cords, cables, window stop, and parting bead can all be removed without care and discarded.

I install replacement windows from the inside of the wall, but first I check that there's enough room between the window stool and the blind stop for the thickness of the new frame. If the space is too tight, I cut the window stool back with a paneling saw or a Roto-Zip®, using a straightedge as a guide *(Figure 5-20)*. If the space is too wide, I fill it later with a small piece of trim on top of the stool.

Before installing the window, I scrape the old jamb clean and apply a bead of caulk at the interior corner of the blind stop *(Figure 5-21)*. The bottom edge of the new frame is installed first, then the top tilted in against the blind stop. Pan-head screws can be used to secure the frame — about every 12 in. — but I prefer to drill holes for flathead screws with a tapered countersink. Then I cover the screw heads *(Figure 5-22)* with an adhesive screw cap.

Chapter 6

Interior Doors

Back in the early 1980s, apartment and condominium construction fueled a building boom in my area. I remember standing with my brother in the middle of a muddy, 380-unit project and staring at receding rows of brand new, two-story townhouses. We had just landed the finish contract — including all the doors, trim, and hardware. We wondered how we would ever make a living installing interior prehungs at $6 a door and $5 a lockset.

But like most carpenters, we jumped into the job feet first. It didn't take us long to learn new tricks for setting prehung doors and installing locksets. By the time we finished that development, we were convinced that production techniques could also save us time on custom homes — our real bread and butter. After all, production carpentry isn't about cutting corners or skipping important details. Production carpentry is about being organized — using the same steps in the same sequence each and every time.

ORDERING PREHUNG INTERIOR DOORS

While interior prehung doors might seem easier to order than exterior doors, it's surprising how often carpenters forget important specs when ordering. Using a standard form eliminates many of these errors and omissions (see the chart, "Interior Door Takeoff," Appendix 2). Obviously, the door's size is crucial, but the jamb width is important, too. The jamb width should match the thickness of the finish wall, including any drywall or sheathing. If the jambs are the correct width, installing the casings is a simple task.

The type of door must be specified, too, whether it's smooth hardboard, embossed, solid core, fir, or paneled. So must the door stop — whether it's flat or molded, $3/8 \times 1 1/4$ in. or $1/2 \times 1 1/2$ in.

Most prehung doors are prebored for locksets. While the standard $2 1/8$-in. face bore matches the diameter of most locks, some deadbolts and many upper-end locksets require smaller bores. Lock strikes vary, too: To make sure the jamb will be mortised properly, specify whether the strike will be a D-shaped lip-strike, or a T-shaped T-strike.

You can save time and on-site labor expenses by ordering prehungs undercut to accommodate the finish floor covering. Most manufactured jambs measure $80 1/2$ in. from the inside of the jamb head to the bottom of the jamb leg. The head gap between the top of the door and the jamb takes up $1/8$ in., so a 6'-8" door usually clears the subfloor by $3/8$ in. That leaves barely enough room for vinyl floor covering with $1/4$-in.-thick underlayment, never mind carpet and pad.

If the floor will be carpeted, find out the thickness of the carpet and pad and have the prehung manufacturer cut the bottoms of the doors accordingly. For most carpets (except tight looms like Berber), I like the doors off the subfloor about $1 3/8$ in., so I order the doors undercut 1 in. For doors over tight-loom carpet or vinyl, I specify the appropriate undercut.

INSTALLING HOLLOW-CORE PREHUNG DOORS

I use two different techniques for installing prehung doors — one for light, hollow-core, 6'-8" doors, and another for larger doors, pairs of doors, and heavy solid-core doors.

Figure 6-1. *Before installing a hollow-core prehung, case the hinge side of the jamb. Glue and nail the miters, then carefully move the door to the opening.*

Figure 6-2. *When adjusting a hollow-core door frame, use a pry bar as an extra pair of hands: Step on the bar to remove the weight of the door from the jamb, and twist the bar to shift the door from one side to the other.*

Contrary to tradition, I don't shim hollow-core doors. I've installed thousands of hollow-core 6'8" prehung doors without shims, and I've never had even one settle by itself. In fact, I once had to return to an apartment building to reverse the swings of six bathroom doors. (There wasn't enough room in those bathrooms for a toilet and a door.) When we trimmed the building two months earlier, we must have installed one of the doors just before quitting time because none of the casing had been nailed to the wall, and the door jamb hadn't been nailed to the trimmers. In spite of these omissions, the door still fit fine. The painter's caulk even made it tough for me to remove the jamb.

Every Door Near its Opening

Before my crew begins to install new doors on a job, I walk the house and mark an X on the wall beside each door opening. (The X will be covered later by the casing.) The X marks the location of the hinges, so the crew will know which way each door swings. Next, we place each door beside its opening, and my crew checks each opening to be sure it's 2 in. taller and wider than the door. (A rough opening for a 3'x6'8" prehung door should measure 38 in. wide and 82 in. tall). Finally, I double-check the entire house personally. It's a lot easier to make changes and move the doors before they're nailed in place.

Hinge-Side First

At one time, hollow-core prehungs arrived on my job sites with the casing attached to the hinge side. But today, manufacturers ship most doors without casing to minimize damage to the trim. While you could install the door and then nail on the casings, there are good reasons for casing the hinge side of the door first *(Figure 6-1)*: It's easier to set the unit with the casing attached; the casing firms up the somewhat weak 11/16-in. jamb stock; and the casing helps straighten out the hinge, head, and strike gaps. The casings don't interfere with shimming the jamb because I never shim hollow-core prehungs.

Use Your Foot

After attaching the casing, place the unit partially in the opening and pull the plug or the duplex nail that temporarily holds the door in the jamb. Once the plug has been pulled, most doors sag towards the floor, widening the gap between the top of the door and the jamb (head gap) on the strike side. It

Figure 6-3. Keeping one foot on the pry bar, you can check for plumb in hollow-core prehungs by using a good long level (far left). If walls, doors, or windows are nearby, however, check for plumb by measuring the margin between the new door and adjacent walls or casings (left).

might seem that you need four hands to straighten the jamb and door in the opening, but I find that a well-placed pry bar on the floor is the only extra help needed *(Figure 6-2)*.

Place the pry bar under the door a bit closer to the lock stile than the hinge stile. If the door has been undercut for carpet, then set a small block of wood on the floor first, so that the tail end of the pry bar is an inch or so off the floor.

Step on the pry bar slightly—just enough to balance the weight of the door—then rack the jamb and door back towards the hinge-side. The head gap will begin to close as the door approaches plumb. I also rotate the pry bar a little towards the strike jamb, which moves the bottom of the hinge jamb away from the trimmer and helps plumb the door, too. Be careful not to step on the pry bar too hard or the door will come off the ground. Find a comfortable stance and apply just enough pressure to hold the door still.

Setting Prehungs Out-of-Plumb

If a door is standing alone, without another door or wall nearby, then use a good level to set the door perfectly plumb and level in the opening. However, if there are nearby walls, doors, and windows, the door casings should line up with them *(Figure 6-3)*. That's why, while I'm stepping on the pry bar, I pull my tape measure to check the margin between the adjacent door or wall at the top and bottom of the casing. I look at the tops of adjoining casings, too.

To achieve even vertical margins between new doors and existing elements, it's often necessary to set prehungs slightly out of plumb. Aligning the top of the new casing with the tops of existing door and window casings occasionally means cutting a jamb down or shimming it up off the floor. Before driving any nails, check the gap between the door and the jamb on all three sides. It should be uniform all the way around, and about the thickness of a nickel. If the floor is out of level and the head gap is too big, then the hinge jamb may have to be shimmed off the floor just a little (more on that in a minute).

The Right Nailing Sequence

I always place the first nail about 2 in. below the miter on the hinge-side casing *(Figure 6-4)*. I shoot the second nail about 2 in. below the bottom hinge. If you don't have a nail gun, slide two oversized shims under the door while driving these first nails.

While these two nails will hold the door still, the jamb will still move easily. This is where I step back a little and take a long look at the head gap and the margins between the casing and adjacent walls or doors. If the head gap is too tight, I lift the strike leg and make sure there's enough wiggle room between the head jamb and the header, so that I can correct the gap later. But I don't nail the strike leg just yet.

If the head gap is too big because the floor is out of level, I place the pry bar beneath the door

Figure 6-4. On a hollow-core prehung, the first nail is driven about 2 in. below the miter on the hinge-side casing (far left), and the second is driven 2 in. below the bottom hinge (left).

Figure 6-5. If the hinge jamb must be lifted off the floor, be sure to install a shim to stop the door and jamb from settling.

on the hinge side and raise the door and jamb straight up until the head gap is about the thickness of a nickel. (With only a few nails in the casing, it's easy to lift the door and the jamb together.) Then I drive a shim between the jamb and the floor so the door won't settle prematurely and crack the miter open *(Figure 6-5).* If the door has to be raised too much, the head casing might not line up with adjacent windows or doors. In that case, the door will have to be removed and the offending jamb leg trimmed down.

The Third Nail

Most doorjambs today are relatively weak — often only 11/16 in. thick. Because of this, the weight and pull of the door can deflect the jamb, causing a "belly" or "tweak" near the top hinge. A bellied or tweaked jamb is more common on heavy, solid-core doors, but it can happen on hollow-core doors too. Deflection of the jamb at the top hinge causes a wide gap to open between the door and jamb above and below the top hinge *(Figure 6-6)*. And if this hinge gap is too big, you can bet the strike gap will be too tight. Here's how I quickly correct jamb deflection.

With the jamb secured by the first two nails only, I place a pry bar under the leading or strike edge of the door. I step lightly on the pry bar, applying just enough pressure to take the door's weight off the jamb and force the jamb and the top

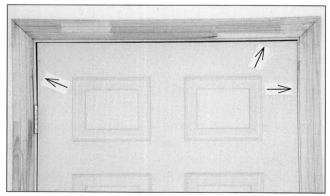

Figure 6-6. Deflection opens a gap above and below the top hinge (left arrow), and causes the door to bind against the strike jamb (right arrow). Deflection also causes the head gap to swell (upper arrow).

Figure 6-7. *If the top of the door isn't touching the stop (left), press the hinge-jamb with your shoe, then fire a nail to secure the jamb. If the bottom of the door isn't touching the stop (right), use your foot to press the strike-jamb before nailing it.*

hinge back toward the trimmer. When I see a gap about the thickness of a dime between the door and the jamb above the top hinge, as well as between the top of the door and the head jamb on the strike side, I drive the third nail through the casing and into the trimmer right behind the top hinge. When I lift my foot off the pry bar, the door will settle slightly and both the hinge gap and the head gap will widen out perfectly. To keep it that way, I shoot a fourth nail an inch or two from nail number three.

The Remaining Casing

The remainder of the jamb can now be fastened to the wall. I drive nails through the casing behind the middle and bottom hinges, as well as through the casing above the head jamb on the strike side, about 3 in. away from the miter. I then tack the strike-side casing to the wall every 14- to 16-in., checking that the margin remains about the thickness of a nickel.

Correcting Cross-Leg Jambs

Before casing the opposite side of the door, I always straighten out any "cross-leg" in the jamb. If the jamb is cross-legged, the door won't lie flat against the strike-side stop.

If the top of the door hits the stop but the bottom won't, just tap the bottom of the strike leg toward the door. If the bottom of the door touches the strike-side stop but the top doesn't, then tap the bottom of the hinge jamb toward the door. I press against the jamb leg with my shoe a little, then smack it lightly with my nail gun *(Figure 6-7)*. Once I see that the door is flat against the stop, I shoot a nail through the face of the jamb, a few inches from the floor.

I always adjust for cross-leg at the bottom of the jamb, not at the top. I like to keep the jamb head flush with the drywall, so it's easier to join the miters on the casing. If the jamb is short of the wall at the bottom of the leg, I flatten the drywall with a hammer before installing the casing.

After applying casing to the stop side of the jambs, but before nailing casing to the trimmers, I check the fit of the door again. If necessary, the jamb can still be adjusted minutely with a hammer, a block of wood, and a few light taps. When I'm satisfied with the fit, I nail off the casing, driving nails about every 14- to 16-in. I also place one nail above the lockset strike and one below it. In addition, I drive a nail through the casing about two inches above the floor — if the baseboard is cut and installed too tight, the nail will prevent it from

Figure 6-8. *Replacing one short hinge screw with a long screw is enough to prevent a hollow-core door from settling.*

Figure 6-9. *When hanging solid-core doors, check that the floor is level across the opening before setting the jamb. Shim the level until the bubbles are perfectly centered, then secure that shim right next to the trimmer where the jamb leg will rest.*

pushing the jamb and ruining the fit of the door. I also shoot a nail through the jamb, above and below each hinge, to prevent the door from settling. Finally, because I never shim hollow-core prehung doors, I replace one of the short hinge screws in the top hinge with a long screw that reaches through the jamb to the trimmer *(Figure 6-8)*. That long screw is enough to prevent any possible settling. I tighten that screw only enough to seat it snugly. Too much torque will draw the hinge and jamb back towards the trimmer and ruin the fit of the door.

INSTALLING SOLID-CORE PREHUNGS

I may not shim lightweight hollow-core prehungs, but I certainly do install shims on all solid-core doors, all pairs of doors, and any door over 6'8" tall. The trick is to shim the floor and the trimmers before slipping the jamb into the rough opening.

Floor Prep First

When I approach a rough opening for a pair of doors or a solid-core prehung, the first thing I do is level the opening at the floor.

I lay an appropriately sized level across the floor and shim the level until the bubbles are perfectly centered *(Figure 6-9)*, placing the shims at the low point of the floor, right next to the trimmer. Then I break the shim off and nail it to the floor. If the floor is concrete, I use caulk to hold the shim in place. Since the bottom of the door jamb will rest on this shim, I no longer have to worry about holding a level over my head while setting the jamb.

Some floors are so badly out of level that one leg of the jamb needs to be cut off in order to maintain the alignment of head jambs and casings. While the level is lying across the opening, measure the thickness of the shim needed to center the bubbles. That amount can be cut off the high leg of the jamb. Rather than cutting the entire amount off one leg, I often split the difference, cutting half the thickness of the shim from one jamb leg and installing a smaller shim beside the opposite trimmer.

Shim the Trimmers

Part of the challenge of carpentry is developing systems for managing large, awkward, and heavy

Figure 6-10. *Plumb the hinge-side trimmer by nailing shims near the top or bottom, whichever is needed.*

Figure 6-11. *After shimming the hinge-side trimmer, measure the opening again to be sure the frame will still fit.*

objects. If you reduce the initial movement of a door frame in a rough opening, you can speed up installation time and reduce job-site frustration. That's why I like jambs to slip into rough openings with only 1/4 in. to spare (that's only a 1/8-in. gap on each side).

To achieve this fit, check the rough opening width and shim the trimmers plumb until the opening is about 1/4 in. wider than the outside dimensions of the new frame. The type, size, and shape of shims can affect the speed and accuracy of your work. I cut my own shims so they're all identical (see "Making Perfect Shims," next page). The trimmer on the hinge side is the most important, but before applying any shims, check that the opening is properly centered in the wall. For example, if the opening is at the end of a hallway, be sure that the wall space on each side of the rough opening is equal. Otherwise it will be difficult or impossible to center the door in the wall.

As with setting hollow-core prehungs, measure to nearby walls or door and window casings. Make sure that the margins between the new door casing and nearby door and window casings or walls will be even. Nail the shims to the trimmers near the top of the jamb and/or near the floor *(Figure 6-10)*.

Hold off shimming near the hinge locations until the jamb is in the opening. The shims behind the hinges need to be adjustable, so it's best to install them after the jamb is in. Before attempting to install the jamb, double-check that the rough opening is still large enough for the door frame to fit in with ease *(Figure 6-11)*.

Plumb and Level

Remove the hinge pins and separate the door from the jamb. Then slide the jamb into the opening. Because the trimmers were shimmed plumb, the hinge jamb can be tacked at the top and bottom immediately. Check that the head of the jamb is level; it usually is if the floor has been shimmed. Then place a level against the hinge jamb.

Stabila manufactures a magnetic level that sticks to the hinge leaves, keeping your hands free for simultaneous shimming and nailing. Master Level offers a spring-loaded level that can be inserted in the opening and also allows for hands-free installation.

Shims should be installed behind every hinge, but start by securing the top of the jamb with a shim behind each leg, one on each side of the head

Making Perfect Shims

I use builders' shims often, but that doesn't mean I like them. They're never the right size or shape, and they're difficult to cut off.

Instead, I prefer to cut end-grain shims from 2x6 scrap wood on my table saw. These homemade shims are always the same size and shape and they're only 1/4 in. thick, so I can slip one behind most jambs, butt end first, then follow it up with the point of another one. These shims perfectly fill any odd-sized gap between a twisted trimmer and the back of a jamb.

Cutting shims on a table saw can be dangerous, but I've tried to reduce the danger as much as possible by using a shop-made sled **(see photo)**. This template cuts each shim exactly the same size and neither of my hands is ever near the blade. My right hand stays firmly on the push block, which keeps it far above and to the right of the spinning blade. I never cut shims from a block smaller than 6 in. long, so my left hand is always at least that far from the blade. I stand back from the blade and slightly to the left side so that any chips blow past my right side.

A shop-made sled makes cutting shims safer and more accurate.

Figure 6-12. *Secure the head of the jamb by placing shims on the hinge side and the strike side. I stick a Stabila magnetic level to the hinge leaves, which frees both of my hands for shimming and nailing.*

(Figure 6-12). Next shim all the hinges. I shoot a nail or two through the jamb beneath these shims to hold them in place, but I never nail through them. It seems that every time I nail through a shim I have to move it later — to make fine adjustments in the fit of the door — which is nearly impossible to do. Though I shimmed the trimmer before installing the jamb, additional shimming, especially behind the hinges, is required to perfect the fit of the door and to secure the door and jamb.

I've found that it's faster to finish shimming, adjusting, and nailing off the jamb with the door swinging on the hinges, so I don't shim or fasten the strike side yet. I nail the strike leg once or twice near the top and the bottom, but leave it mostly dangling. Before attempting to swing the door, I measure the opening and make sure it's wide enough so that the door will clear the strike leg.

A Long Hinge Screw

Before hanging the door on the hinges, there's still one more important step: I replace one of the hinge screws in the top hinge with a screw long enough to reach through the jamb and penetrate at least an inch into the trimmer. Heavy, solid-core doors are more than a match for the 15-ga. finish

Figure 6-14. *Be sure to shim behind the lock strike, too, and fasten the jamb securely to the trimmers at this weak point.*

Figure 6-13. *A long screw can also be placed behind the hinge, but only after the door is installed (top). Therefore, during installation use an oversized shim to help remove the weight of the door from the top hinge (above).*

nails I shoot, and screws are the only answer for permanently securing a jamb.

Sometimes it's hard to come up with a screw that's the right size or the right color to match the hinge. When that's the case, install a screw behind the hinge. But before burying a screw behind a hinge, hang the door on the hinges and complete most of the installation. To prevent the weight of the door from pulling on the jamb, lift the door with a pry bar and use an oversized shim to hold the door while adjusting and nailing the jamb to the trimmers *(Figure 6-13, top)*.

Oversized shims are also handy for swinging and adjusting pairs of doors. Once the hinge leg is adjusted properly, remove the top hinge leaf from the jamb, countersink a hole in the hinge mortise, and install a long screw that reaches well into the trimmer *(Figure 6-13, bottom)*.

Seating the long screw, whether it's behind the hinge or through the hinge, must be done carefully. Too much torque will ruin the fit of the door by changing the width of the gaps around it. Check the gaps between the head of the door and the jamb, above the top hinge and at the top of the strike stile, before shimming and fastening the remainder of the strike jamb. The long hinge-screw can be tightened to increase the strike gap, or it can be loosened to decrease the strike gap.

Check for Cross-Leg

Before securing the strike leg, be sure that the jamb isn't cross-legged. The door must lie flat against the stop on the strike side of the jamb (see "Correcting Cross-Leg Jambs," page 83). The bottom of the strike leg is easy to adjust, and moving the bottom of the hinge leg is still possible too (only the top hinge is screwed to the trimmer). Once the door is

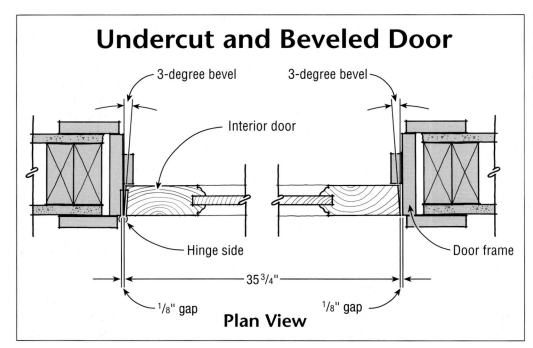

Figure 6-15. *Undercut and beveled (UB) doors are manufactured 3/16 in. under nominal size, and both stiles are beveled three degrees. Originally introduced for prehung manufacturers, they also work well shipped right to the job site.*

lying flat against the door stop, shim the strike jamb so that the gaps are perfect.

Shim Behind the Lock Strike

Always shim behind the lock strike to strengthen the jamb at one of its weakest points *(Figure 6-14)*. But also shim the strike jamb near the floor to protect the jamb from being kicked out of position by an errant foot or vacuum cleaner. On tall doors I install an additional shim halfway between the lock strike and the top of the door (remember, the very top of the jamb was shimmed earlier).

Once the jamb is shimmed and fastened, break off the ends of any projecting shims and case both sides of the opening. On solid-core doors over 6'-8" — after the casing is complete — I place a long screw through the second hinge, too, as a further precaution against settling.

NEW DOORS IN NEW JAMBS

For most solid-core interior doors, especially if they're over 6'-8" tall, I prefer installing slab doors rather than prehungs. Tall solid-core prehungs are unwieldy, and they're easily damaged during transportation. (Carpenters are often damaged transporting them, too.) Fortunately, slab doors aren't nearly as difficult to hang on the job site as they once were.

Undercut-and-beveled doors (we call them UB) exemplify how new products often challenge accepted installation techniques and beg for improved carpentry routines *(Figure 6-15)*. These new slab doors prove that job-site productivity can be increased without sacrificing quality. Unlike full-sized slab doors, UB doors are manufactured 3/16 in. under nominal size: A 2'-8" undercut door measures 31 13/16 in. rather than 32 in., and both stiles are beveled three degrees. These doors were originally introduced for prehung manufacturers, but they work well shipped straight to the job site, too.

Set the Jamb

UB doors eliminate the worry about ordering right-hand and left-hand doors in advance and need only to be hinged, bored, and trimmed for height on the job site. But setting a jamb for a UB door requires a slightly different attitude than setting a jamb for a full-sized slab door; setting the jambs and hanging UB doors can be accomplished in far less time than it takes to hang a true slab door.

I set interior jambs for UB doors almost the same way I set solid-core prehung jambs. The floor and the trimmers should be shimmed first, so that the jamb legs and head will go in plumb and level right from the start. I tack the jamb in the opening (see the previous section "Installing Solid-Core Prehungs") with a few nails, then use a long level to check that the jamb is plumb. My long levels are marked with tape that corresponds to the hinge locations on my router templates, which enables me to install shims near the hinge locations *(Figure 6-16)*.

As with solid-core prehung jambs, never nail

through the shims. Secure the jamb by nailing beneath the shims, so that the jamb can be adjusted after the door is installed. It isn't necessary to shim or permanently secure the strike leg of the jamb. Instead, tack the strike jamb to the trimmer with three or four nails. That way, once the door is swinging, the jamb can be adjusted to fit the door.

While I'm at the opening, I measure the height of the jamb and subtract the thickness of the finished floor covering so that I can cut off the bottom of the door at my bench.

Hinge the Door

After the bottom of the door is cut for finished flooring, I set the door on edge in my bench with the hinge-stile up. To prevent errors, I always use the same systematic techniques when I work at my door bench (see Chapter 4, "Hanging a New Door in an Old Jamb," page 58). I always pretend that the bench is on the stop side of the door, and the hinge barrels always point toward my belly. The bevels are always falling in toward my bench. And when I work on the hinge stile of a door, the top of a left-hand door is always on my left; the top of a right-hand door is always on my right *(Figure 6-17)*. Period. I use the butt-in-the-butts method for identifying door handing. (See Chapter 4, page 45.)

Full-length hinge templates are relatively inexpensive, and they're essential for production-hang-

Figure 6-16. *Use blue masking tape to transfer the hinge locations from your template to your jamb level so that all shims will be perfectly placed.*

Figure 6-17. *If the top of the door is in the hook, then it's a left-hand door (butt-in-the-butts). If the bottom of the door is in the hook, then it's a right-hand door (butt-in-the-butts).*

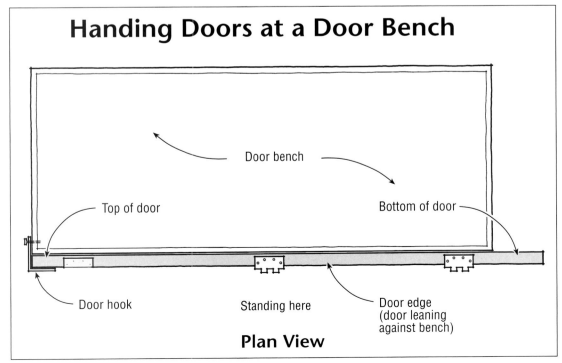

INTERIOR DOORS 89

Figure 6-18. *Before attaching the template, adjust the backset button for the thickness of the door. In this example, the backset button is adjusted for a 1 3/8-in. door.*

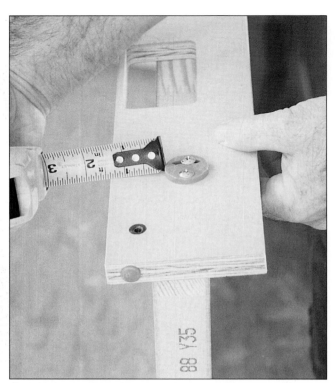

Figure 6-19. *Once the top of the template (not the spacer button) is flush with the top of the door, drive in the template nails.*

ing new doors in new jambs (see "Full-Length Hinge Templates"). I attach the template flush with the top of the door and ignore the head-gap button for now — more on that button below *(Figure 6-18)*.

The inside narrow edge of the template should be flush with or slightly proud of the door on the stop side. Backset stops, located on the underside of the template, make it easy to mortise the door and the jamb with identical hinge backsets, so that every door will be flush with the edge of every jamb. The backset stops must be adjusted properly, but that's an easy task: While the template is attached to the door, simply loosen the backset buttons and rotate them until they just touch the door, then tighten them securely *(Figure 6-19)*.

Once the template is correctly attached, routing multiple mortises is quick work. After installing the hinges, I rotate the door in my bench, bore for the lockset, and mortise for the lock-latch plate. Working methodically, I can usually process (cut the bottom of the door, hinge, and bore) an undercut-and-beveled door in about fifteen minutes.

Hinge the Jamb

Before hanging the door, the jamb needs to be mortised for the hinges, too. That job is made especially easy by the head-gap button on the top of the template. When the template is attached to the jamb, the button acts as a spacer, creating a perfect head-gap clearance of just under 1/8 in. *(Figure 6-20)*.

To attach the template, place it against the jamb with the backset stops touching the edge of the jamb on the hinge side. Slide the template up until the button gently hits the head jamb, then drive in the top template nail. Always check that the backset stops are touching the jamb before driving in more template nails. I rout the jamb with one hand and use my other hand or one foot to hold the template tightly against the jamb *(Figure 6-21)*, just in case one of those little nails comes loose. That's happened to me more than once.

Adjust the Jamb

The steps for finishing the job are the same as for installing a solid-core prehung. Close the door and look at all the gaps before adjusting the jamb. Some adjustments are usually necessary, but if the hinge jamb was set plumb and the head jamb level, those adjustments should be minimal.

Several adjustments affect the head gap. The weight of a heavy 8-ft. door not only will pull on the jamb at the top hinge (which causes a wide

Full-Length Hinge Templates

Single-hinge templates are great for hanging new doors in old jambs, and they can be used for new doors in new jambs too. But don't attempt to save the cost of a full-size multi-hinge template, especially if you're installing 20 or 30 doors. Once purchased and reasonably cared for, full-sized laminated templates last indefinitely and quickly recoup their initial cost: 6'-8" templates cost less then $50 each, and 8-ft. templates run just over $50 a piece (see Appendix 1).

A full set of templates can be collected over time, as they're needed, and should eventually include two 6'-8" templates, one for 3 1/2-in. hinges (for 1 3/8-in.-thick doors), and one for 4-in. hinges (for 1 3/4-in.-thick doors). Two 8-ft. templates are also useful, one for 3 1/2-in. hinges and one for 4-in. hinges. Templaco's 8-ft. templates are equipped for the installation of three or four hinges per door **(Figure A)**.

For taller doors and custom-size doors, I use a Bosch Adjustable Hinge Template (Model 83038) along with Bosch's four-hinge conversion kit (Model 83039). I use an adjustable template whenever I have to match existing hinge mortises for more than two doors. I also use my adjustable template for windows and custom-sized doors, and whenever I hang wooden doors in steel jambs. I prefer the Bosch design because it's simple to set up and incorporates a swivel-stop that allows instant changes between right-hand and left-hand doors **(Figure B)**.

Figure A. *On production jobs, a full set of hinge templates is a must.*

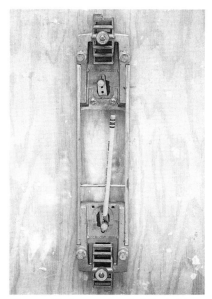

Figure B. *I use a Bosch Adjustable Template for tall and custom-sized doors.*

hinge gap above the top hinge and a wide head gap above the lock stile), but will also compress the bottom of the jamb and the bottom hinge (which can also cause a wide gap above the lock stile). Therefore, look at the entire jamb before beginning any adjustments, and never attempt to correct the fit of the door with a single adjustment.

Install the Door Stop

For a beginner, door stop might seem like the easiest thing in the world to install, but anyone with experience knows that there's a right way and several wrong ways to do the job. First, be sure to fasten the door stop to the jamb before applying the casing, just in case the jamb moves a little while nailing on the door stop (without the casing attached, the jamb can still be adjusted easily with a pry bar or a few shims).

Figure 6-20. *The spacer button holds the template off the head jamb and guarantees a perfect head gap on every door.*

Figure 6-21. *Use a full-length hinge template on new jambs. Template pins secure the template to the jamb, but a well-placed foot ensures that the template won't move.*

Figure 6-22. *On the hinge side of the jamb, hold the stop away from the door about the thickness of a dime, so that the door will never bind on the stop.*

Miter the door stop where the legs meet the head, and cut the legs long enough to reach the bottom of the jamb. After the door stop is applied to the jamb, the face of the door should ideally be flush with the jamb. However, the ideal is sometimes not possible: Some jambs, especially old ones, can be slightly bowed or cross-legged, leaving the door flush with the jam at the top, but proud of the jamb at the bottom. If the jamb can't be moved, try to split the difference: Apply the doorstop so that the jamb is just a little proud of the door at the top, and so that the door is slightly proud of the jamb at the bottom.

Once I've determined if the door will be flush with the jamb, or the jamb slightly proud at the top of the door, it's time to apply the stops. When I'm working alone, I make a pencil mark on the head of the jamb where I want the inside of the door to stop — for a 1 3/8-in.-thick door, I measure in 1 3/8 in. from the face of the jamb — a little more if the jamb must be slightly proud of the door. If another carpenter or helper is nearby, I skip the measurement business. Instead, I ask that person to stand on the hinge side and hold the door flush with the jamb while I shoot one nail through the head. I install the head first, and secure it with one nail about 2 in. from the strike side of the jamb. After driving that first nail, I check that the door is flush with the jamb when it hits the stop.

The door should touch the stop on the strike side of the jamb, but it must not do so on the hinge side; otherwise the door will pinch the stop as it closes (in which case we say the door is stop-bound).

Before nailing the head stop, I pull it away from the door on the hinge side about the thickness of a dime, then drive one nail about 2 in. from the miter and one or two nails in the center *(Figure 6-22)*.

Once the head stop is fastened to the jamb, attaching the stop legs is quick work. Start with the stop on the strike side. Join the miter tightly and hold the leg straight, then fasten it to the jamb with one nail about 2 in. below the miter *(Figure 6-23)*. Next, close the door and hold it gently against the head stop, press the dangling leg snugly against the door, and fasten it every 12- to 14-in. *(Figure 6-24)*.

The stop leg on the strike jamb should touch the door evenly, from the top of the jamb to the bottom. However, the stop leg on the hinge side of the jamb must follow the head stop and be held

Figure 6-23. *Once the head is installed, the legs are easy. Start at the top by securing the miter, then close the door against the head stop.*

Figure 6-24. *Hold the door gently closed while fastening the leg to the jamb every 12 in. or so.*

Figure 6-25. *Space the stop on the hinge jamb away from the face of the door about the thickness of a dime, so that the door won't bind on the stop.*

away from the door evenly, from top to bottom, about the thickness of a dime *(Figure 6-25)*.

METAL JAMBS AND PREFIT DOORS

Most residential contractors don't install metal jambs often enough to get very good with them. But metal jambs and prefit doors are becoming more and more popular — and although they can have sharp edges, they don't bite. A wide variety of aluminum and metal frames are marketed in my area, and I often install them on tenant-improvement (T.I.) jobs in commercial buildings. But I've installed these frames for both security and fireproofing in custom homes, too. Trust me, it's a lot harder to order metal jambs and prefit doors correctly than it is to install them; and if they're ordered right, they go in fast.

Ordering Metal Jambs

Before ordering metal jambs, you should be armed with three important pieces of information: the opening size, the swing, and the type of hardware. First, be certain to order the jamb for the proper wall width. A metal jamb is similar to a wood jamb, but the sizes are usually different. In my area, I order a stock 4⁷/₈-in. metal jamb in place of a standard 4³/₄-in. wood jamb, or a 5³/₈-in. metal jamb rather than a 5¹/₄-in. wood jamb. (The metal jamb industry seems to have accepted the reality that the sum of stud width and drywall thickness is greater than their ideal measurements; by contrast, the wood jamb industry still seems to believe that a 2x4 wall with ⁵/₈-in. drywall on each side measures exactly 4³/₄ in.)

Before framing, check with your metal-jamb manufacturer for recommended rough opening sizes. Those sizes vary from one manufacturer to the next: For instance, Timely Industries manufactures metal jambs that require unusual rough openings that must be 1¹/₄ in. wider and 1 in. taller than the door.

Second, order the correct swing and discuss the handing thoroughly: Different manufacturers use different methods for handing a door. Be sure you explain to your supplier how you hand your doors. If necessary, draw a diagram and fax it along with your order. Some manufacturers use the "back-to-the-butts method" — with your back against the hinges, the handing of the door is the same as the swing of your arm. Other manufacturers use the hinge-side method — facing the hinge side of the door, if the hinges are on the right, the opening is considered a right-hand door. (On one job I ordered jambs from one manufacturer and doors from another, and was surprised to learn that each company used a different handing method. Fortunately, the numbers were almost even and I ate only two doors, but that was more than my appetite.)

Third, choose the lockset and deadbolt before ordering the door and jamb, and include that information, along with the hinge specifications (the size

Figure 6-26. On large jobs with metal jambs and prefit doors, it's a good idea to install all hardware at a central location, then wheel doors to their openings.

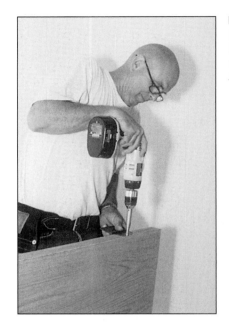

Figure 6-27. Drilling pilot holes for all hinge screws will prevent the door stile from splitting.

and thickness of the hinges), with the prefit order. Most manufacturers will prepare a jamb and door for almost anything, from mortise locks to flush bolts — I even have prefits prepped for panic hardware.

Scattering the Jambs and Doors

Though jambs are prefit to their doors and the doors are prefit to the jambs, the doors are not delivered "hanging" in the jambs like hollow-core interior doors are, which makes it crucial to match the right door to the right jamb. On a large job, this is a matter of checking and double-checking everything.

Once the doors and jambs are on the site — and before installing anything — I place every jamb near its opening. Having several different jamb sizes and door swings on one job is common, along with different hour ratings and hardware preparations. Then I check every opening, sometimes twice. It's easier to move the jambs around before they're screwed into the trimmers.

I do the same thing with the doors. On a large job, I like to install all the hinges on the doors in one central location, using an electric drill and screw shooter, then wheel each door to its correct opening. There, I double-check the door size and swing, and match the door to the jamb at the same time *(Figure 6-26)*. I use a Rok Buk® door dolly (see Appendix 1) to wheel doors around. On a small job, like this one, I hinge the doors with cordless tools while they're standing in the dolly next to the opening.

To prevent splitting the door stiles when attaching hinges, always drill a pilot hole for every hinge screw *(Figure 6-27)*. A Vix® bit centers each hole perfectly, speeds the process dramatically, and ensures that the screws won't move the hinge. To speed production further, I use two cordless drills, one for the Vix® bit and one for my magnetic bit holder.

Installing the Jamb

Many companies manufacture metal jambs. Their designs vary considerably, but most installations begin with the head jamb. Be sure the door rabbet is facing the correct direction, then angle the jamb up and slip it over the drywall at the header *(Figure 6-28)*. Push it up as far as it will go. Next, install one of the legs. It doesn't matter which leg you install first, just be certain to install the hinge jamb on the hinge side! After checking the door rabbet again, angle the upper end of the leg up toward the corner of the header, then straighten it vertically and slip it over the drywall *(Figure 6-29)*. Repeat the procedure for the opposite leg.

Once the jamb is "kicked in," pull the head down onto the tops of the legs, and check that the legs are seated firmly on the floor. Metal jambs have different types of corner-fastening systems. Some jambs have alignment tabs. The tabs help lock the head jamb into the jamb legs. Use an awl or screwdriver to pull the head down tightly against each leg *(Figure 6-30)*. If the jamb has a dif-

Figure 6-28. *Metal jamb installation begins by slipping the head jamb over the header.*

Figure 6-29. *After the head jamb has been installed, angle the upper end of one leg toward the header, then straighten it and slip it over the drywall.*

Figure 6-30. *Use an awl to pull the head down and lock it onto each leg of the metal jamb.*

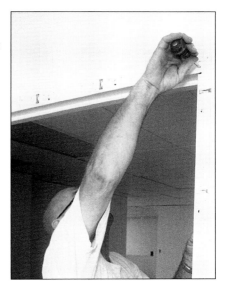

ferent fastening system, like screw-on corner clips, install the clips now.

Swinging the Door

I'm usually alone when I install prefit doors, even 8-ft.-tall ones, so I've learned an easy way to get the door and hinges on to the jamb. Here's how it works.

Set the door in the opening at a 90-degree angle — in other words, as if it's swung open to 90 degrees. Next, slide the door back until the hinge stile is about centered under the head jamb. Then rock the door up on the strike stile so that the top hinge on the door comes up near the top hinge mortise in the jamb *(Figure 6-31)*.

If the door is too high, give it a light kick with your foot near the bottom, and lower the hinge stile until the hinge lines up with the hinge mortise in the jamb. Then drive in one screw. The door will hang — though a little precariously — in the jamb with the top of the door secured by even one screw. Now place your foot against the bottom of the door and push the door with your toe and one hand until the middle hinge comes up to its hinge mortise. (The other hand is still holding the screw gun, which should be loaded and ready with a screw!) Hold the door still with your foot and secure the middle hinge with one screw *(Figure 6-32)*. It's now safe to let go of the door and install the bottom hinge and remaining screws. If the weight of the door has pushed the bottom hinge past the mortise, use a small pry bar between the jamb and the door to help align the hinge.

Adjusting and Securing the Jamb

I use a small pry bar and a block of wood to adjust the door and jamb. Some larger pry bars, especially the type with the really crooked U-neck, work great for metal jambs and metal or wooden doors. Place the pry bar beneath the door near its center, but a little closer to the strike side than the hinge side. I place a screw in my cordless driver, then step on the pry bar with enough pressure to take the weight of the door off the frame, but not so much that I lift the frame off the floor.

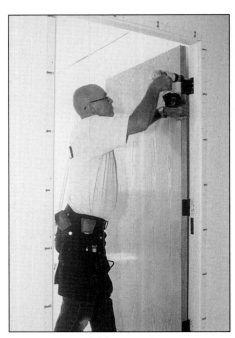

Figure 6-31. To hang a heavy door alone, rock the door up on the strike stile until the top hinge screw lines up with the predrilled hole in the jamb, then drive one screw.

Figure 6-32. Hold the door still with one foot and secure the middle hinge with one screw. Now you can let go of the door and drive the remaining screws.

Usually, I need to push the jamb back toward the hinge trimmer a bit, especially at the top of the jamb because the weight of the door pulls the hinge jamb toward the strike jamb (which also increases the head gap).

While pushing the hinge jamb toward the trimmer, I watch the margin between the jamb and the adjoining wall. Once the margin is parallel, the jamb should be plumb and the head gap should be about the thickness of a nickel. If the head gap is still too big, then the jamb must be raised a little. Whenever I raise a jamb off the subfloor, I always install a shim beneath the jamb so that it won't settle later and ruin the fit of the door. If the margin is too tight, the head jamb can always be raised a little.

If there's no adjoining wall to help judge that the jamb is plumb, I set a level near the opening before beginning the jamb installation. I use the level to check that the jamb is plumb. Once I'm comfortable that the jamb is plumb — and after checking that the head gap is either correct or a little too tight — I drive in the first screw.

The Correct Fastening Sequence

Just like interior prehung doors, the first three screws used to secure the hinge jamb are the most important and must be placed in exact order and location. Hold the door still with one foot on the pry bar, then start by driving a screw into the very top hole in the jamb leg *(Figure 6-33)*. Next, install a screw through one of the holes beneath the bottom hinge. Now check the jamb for plumb again, using a level or by measuring to the adjacent wall. If the jamb isn't plumb, remove one of the screws and correct the problem.

The critical part of installing any metal jamb is the next step. The weight of most doors will cause a small belly or bow in the hinge jamb, right at the top hinge. Naturally, the bow in the jamb will increase the gap between the jamb and the door just above and below the top hinge (the hinge gap). If the hinge gap isn't corrected, the door will be too tight against the strike jamb. The door might even push the strike jamb away from the head jamb, which will ruin the fit of the casing. To correct the problem, press a little harder on the pry bar and force the top hinge and the jamb back towards the trimmer. (If you're having trouble holding the screw gun and driving the screw while stepping on the pry bar and holding the jamb, then lift the door with the pry bar and scoot in an oversized shim.) Once again, I like to push hard enough so that the door just touches the head jamb above the strike stile, and the hinge gap above the top hinge is almost entirely closed. Then I drive a screw behind the top hinge. Once that screw is set securely, I remove my foot from the pry bar. The door will then settle back down slightly, and the hinge gap and head gap will almost always be just right.

Finishing the installation is simple stuff. Install the remainder of the screws in the hinge jamb, dri-

Figure 6-33. *With metal jambs, start by driving a screw in the top-most hole in the hinge jamb. Then drive a second screw below the bottom hinge and check for level.*

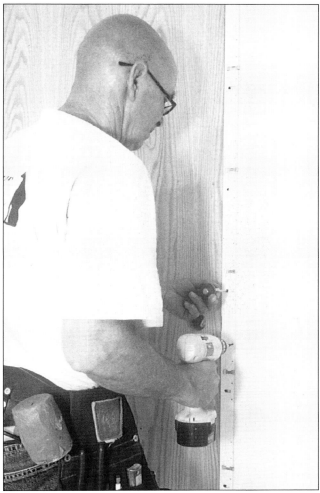

Figure 6-34. *Install two screws near the deadbolt and the lockset.*

ving two screws behind each hinge, plus a screw every 12 in. Then pull the head jamb down tight against the hinge leg and place one screw at the end of the head to lock the two jambs together. Adjust the strike jamb next. Use an awl to pull the strike jamb towards the door until the gap is about 1/8 in. Sometimes the strike jamb must be lifted slightly to correct the head gap. Place a small pry bar beneath the jamb to lift. Once the head gap and strike gap are equal, install a screw in the top of the strike jamb first. Next, use an awl to pull the head jamb down against the top of the strike jamb, so that the joints between the jambs are closed up snugly, then drive a screw through the corner of the head jamb. Install a screw every 12 in. down the strike jamb, using an awl to adjust the strike gap as you go. I like to put two screws through the jamb near the lockset and deadbolt for extra security ***(Figure 6-34)***.

All that's left is installing the casing. As I said before, metal jambs vary from one manufacturer to the next. Each manufacturer has its own fastening system for the casing, but most systems rely on some type of spring clips. To save my valuable hands, I use a rubber mallet to snap on the casing. I cover the mallet with tape so the head won't mar the jamb. But hitting a thin metal casing with a mallet must be done carefully. Used incorrectly, a mallet will dent the casing (I had to replace all the casing on one job because of a hopeless helper). Used correctly, a mallet speeds up the job considerably.

The first piece of casing to install is the head (some casing kits come with corner clips that have to be installed in the head before mounting the casing to the jamb). Center the head over the jamb and begin fastening at the middle of the casing. I apply most of the pressure needed to snap the cas-

Figure 6-35. *Use a rubber mallet to nudge the casing over the clips.*

ing over the spring clips with my left hand, then, at the last moment, I tap the casing lightly with the mallet, taking care to hit it on the edge and nudge it that last little bit *(Figure 6-35)*. The seat is almost always solid. Install the legs last, beginning at the miter and working down to the bottom of the jamb.

Chapter 7

Casing

Like most finish carpenters, I enjoy installing trim more than anything else. Trim molding isn't too heavy (except for large MDF crown moldings). Trim installation doesn't require setting up too many back-breaking tools (often only a chop saw, compressor, and one or two nail guns). And applying trim results in an almost immediate visual reward. Though baseboard is where most carpenters begin their careers (see Chapter 8), door casing must be installed before baseboard, so this chapter appears first.

SIMPLE CASING

Before I roll out my nail gun, I measure, cut, and spread every piece of casing to every opening. That way I don't have to spend a lot of time walking back and forth to the saw and the material pile. If I'm working with another carpenter, one of us will cut and spread material, while the other installs.

To increase both production and accuracy, I cut almost all window and door casings using repetitive flip-stops on my miter saw (see "Making a Repetitive Stop System," page 27). After all, most windows and doors are manufactured in standard widths and heights that repeat throughout a home. For custom sizes, all of my stops flip out of the way so that I can change lengths at any time and still not lose the previous setup.

I cut the longest lengths of casing first, usually beginning with the legs for the door jambs. After

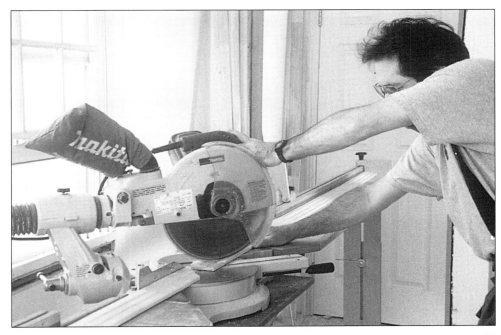

Figure 7-1. Cut a left and a right casing leg for each jamb. Cut two of each if you're cutting casing for both sides of the jamb.

Figure 7-2. *If the miter doesn't close because the drywall is a little proud of the jamb, try sanding off the back of the miter.*

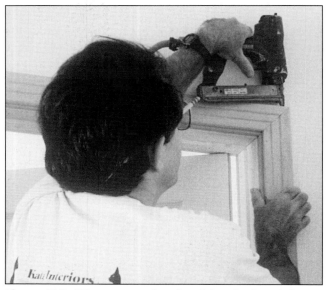

Figure 7-3. *When you get the miter tight, glue it and shoot a brad nail through the miter from both directions (if possible).*

that, I cut the heads, which often requires three flip-stops—one for 2/6 doors, one for 2/8 doors, and one for 3/0 doors. I usually set up stops for window casing, too, again because window sizes almost always repeat throughout a home.

Miters on Casing

Cutting miters on casing is fairly simple for beginners to visualize because the material is mitered flat on the table of the chop saw, exactly as it mounts on the jamb or wall. All casing miters (except self-returns on window aprons) are cut in the same manner, like picture-frame cuts—unlike the ever-changing vertical miters on baseboard.

Whenever I train a beginning carpenter, I like to point out several key concepts about casing.

This molding is almost always cut with the short point of the miter on the thinner, or inside, edge of the molding; the long point of the miter is almost always on the thicker, outside edge of the molding. Furthermore, all measurements are taken and marked on the short point of the miter, so always keep the thicker, outside edge of the casing against the fence. That way, measurement marks are easily visible on the near edge of the casing when you position the casing in line with the saw blade.

Remember, too, that every set of casing on a door has a left leg, a right leg (each is mitered in an opposite direction), and a head. Heads are mitered at both ends, which makes cutting them a no-brainer. But legs are mitered at only one end and they're not reversible. To be certain that I always cut a left leg and a right leg for each jamb, I place two uncut legs on the extension wings of my miter saw, one on each side of the saw blade *(Figure 7-1)*. I cut one with the motor swiveled to the left and one with the motor swiveled to the right. I don't waste time thinking about which one's the left side and which one's the right side; I just like to make sure I have one of each instead of two lefts and no right.

Measuring Door Casing

I precut door casings with a standard 3/16-in. reveal. For the head casings, I measure the inside dimension at the head of the jamb and add 3/8 in.; for the legs I add 1/8 in. so they're off the floor just a little bit and don't have to be cut twice for length. (See "Avoiding a Tape Measure.")

I measure and make a pencil mark on the inside edge of the molding (at the short point), then position the material so that the chop-saw blade enters precisely on the mark. On my Makita saw, I lower the blade until a tooth is just about to touch the material, then line up the mark with the edge of the tooth. On my Hitachi 15-in. saw, I look through the viewing window on the guard housing to line up the mark with the blade. Beginners might want to cut each piece a little bit long at first, then move the material gradually toward the blade until it's cutting right on the mark.

Installing Door Casing

I brad-nail the head casing to the jamb first, so that it's well secured. I've developed a good eye for reveals, so I no longer trace a reveal line on the

Avoiding a Tape Measure

When I'm installing casing on hard-surface finished flooring — tile, stone, or wood — I cut all the legs a little long and re-cut each one in place, so it fits perfectly. I'm a firm believer in avoiding tape measures whenever possible since math mistakes are too common. Instead, I turn the leg upside down and trace a pencil mark from the top of the head casing onto the outside edge of the leg **(see photograph)**. A cordless saw works great for this job because it cuts fast and clean, and doesn't tie my air hose in knots.

On finished floors, cut the door leg casings a little long and mark them in place: Turn the leg upside down and trace a pencil mark from the top of the head casing onto the outside edge of the leg.

---

jamb, but beginners should use a pair of scribes or a reveal block (as I'm using for window casing in Figures 7-15 and 7-16, page 107) to mark an even line all the way around the jamb. Once the head casing is attached, I hold a leg up to the miter. If the drywall is proud of the jamb, even a little bit, the casing won't lie flat, which means the miter won't close tightly. A longer finish nail (2 in. to $2^1/2$ in.) can be driven through the head casing to draw it tighter against the drywall. But that doesn't always solve the problem, and sometimes the casing has to be pried away from the wall later, which means reaching for one more tool.

So, instead of nailing the head casing to the wall at this time with $2^1/2$-in. nails (and creating another problem), I try sanding off the back edge of the leg's miter just a little, which usually allows the miter to close up tightly. For this type of job, and several others, I carry a sanding block in my tool belt. It's just an 80-grit belt from my belt sander that I've stretched tightly around a wooden block *(Figure 7-2)*. Having a wide, flat, hard surface behind the sandpaper helps control the cutting action of the paper. And the sanding block saves tedious trips back to the chop saw.

After checking that both miters fit together precisely, I apply a thin layer of glue to the miter, then brad-nail the leg to the jamb. To lock the miter together, I shoot a brad nail through the miter, too, if possible from each direction *(Figure 7-3)*. Before firing either of those nails, I run my finger over the miter, to clear any glue out of the way and to assure myself that the joint is flush and tight.

Once the entire casing has been brad-nailed to the jamb, I switch nail guns and drive $2^1/4$-in. nails placed toward the outside edge of the casing and into the trimmers or jack studs, permanently securing the casing and the jamb. (See "Narrow Jamb Extensions," next page.)

Scribing Casing

Casing doesn't always fit where it's supposed to fit — sometimes hallways are designed too small, or doorways and windows are framed too close to adjacent walls. Other obstructions often prohibit the installation of full-width casings, too, including cabinets, countertops, wainscoting, or projecting brick and stone. In each of these situations, the casing must be scribed and cut to fit perfectly into an often irregular space.

Scribing casing to fit isn't nearly as difficult or as frustrating as some carpenters think, and the reward can be worth the effort, especially if you get it right the first time — or at least learn to scribe perfectly after the second try. Occasionally carpenters have a difficult time visualizing just what scribing is all about. I know it took years before I understood the simplicity of it. Now I always remember that scribing casing is simply a matter of removing enough material from the outside edge so that the inside edge will align parallel to the jamb, with exactly the right amount of jamb reveal. To achieve a perfect scribe — often

Narrow Jamb Extensions

When it comes to installing casing, one of the most common problems is that jambs are often a bit shy of the drywall.

If the drywall is proud of the jamb by only 1/8 in. or so, it can be beaten back with a hammer, especially across the head, near the miter locations, and part way down the jamb legs. But for frames that are short of the drywall by more than 1/8 in., jamb extensions must be installed.

Jambs are often 1/2 in. short of the drywall because it's a common error on the West Coast to forget about shear wall when ordering jambs. Half-inch extensions are simple, but I still tackle them the way I do every piece of trim on a project — with an eye toward speed and production.

As with most other pieces of trim, I walk the entire job and list all the jamb extensions for every opening. Then I cut everything at once and scatter all the pieces to each jamb.

I cut extension heads long enough to reach past the legs, and the legs should be cut 1/8 in. longer than the height of the jamb to allow for the reveal on the head jamb. To avoid splitting narrow extensions, I just tack them with a brad nailer **(see photograph)** because they'll be fastened more securely by the casing nails.

Avoid splitting narrow jamb extensions for the head and leg by attaching them with a few brads.

Figure 7-4. To scribe a casing, begin by measuring the narrowest point between the jamb and the wall or obstruction.

Figure 7-5. Tack the casing at the narrowest point, and use a square to measure how far the casing overhangs the jamb.

Figure 7-6. Set your scribe to the measured overhang, or set it directly, as here, positioning one leg on the jamb and one on the edge of the casing at the narrowest point. Either way, remember to add the reveal.

on the first try — be sure to start with the casing parallel to the jamb.

I use a tape measure to find the narrowest point between the inside edge of the jamb and the abutting wall or obstruction *(Figure 7-4)*. Sometimes the narrow point is at the top, sometimes it's at the bottom, and sometimes it's in the middle. I tack the casing at the narrowest point and measure how far the molding overhangs the face of the jamb. It's difficult to accurately read a tape measure when the casing is almost an inch away from the numbers, so I use my square to measure the distance between the inside edge of the casing and the face of the jamb *(Figure 7-5)*. Then I position and tack the casing in place so that the overhang is exactly the same everywhere else on the jamb. Two more brads are usually enough to hold things still.

Scribing a piece of casing on the hinge side of the jamb adds a little more challenge to the job. I position the casing on top of the hinges and tack it to the jamb at the narrowest point between the jamb and the wall. An extra brad or two — and sometimes a real nail or two — is usually needed to keep the casing from rocking on the hinge barrels, and to secure it parallel with the jamb.

Once the casing is temporarily secured to the jamb, I reach for my scribes. Sometimes I use my tape measure and spread the scribes apart exactly the same amount that the casing is overhanging the jamb (if I remember that amount). Of course, I then have to add the amount of jamb reveal required. Most of the time I move more quickly and simply spread my scribes apart right next to the casing, with the point touching the jamb and the lead touching the inside edge of the casing. Then I spread the scribes a little more — about 3/16 in. — for the reveal *(Figure 7-6)*. Finally, I hold the scribes perpendicular to the jamb, run the lead along the abutting wall or obstruction, and slowly scribe a sharp pencil line on the face of the casing *(Figure 7-7)*.

I like to cut scribe lines with a high-rpm panel saw because the blade is small and can follow irregular lines *(Figure 7-8)*. A high-rpm saw also allows me to cut slowly, without the blade binding, so that I can follow the line perfectly. I tip the saw slightly, too, so that the cut is beveled — a little back-cutting allows the casing to clear imperfections in the corner.

Some scribes, like that for the countertop in Figure 7-7, can't be cut with a panel saw, so I turn to my jigsaw *(Figure 7-9)*. No matter which tool

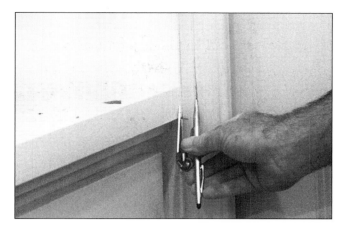

Figure 7-7. *Hold the scribes perpendicular to the jamb and slowly trace the shape of the obstruction (top) or abutting wall (above) onto the casing.*

Figure 7-8. *A small-diameter, high-rpm panel saw works well for cutting to a scribe line.*

Figure 7-9. *For a countertop scribe like this, use a jigsaw.*

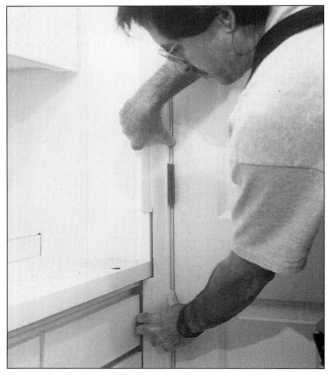

Figure 7-10. *The scribed casing should fit perfectly.*

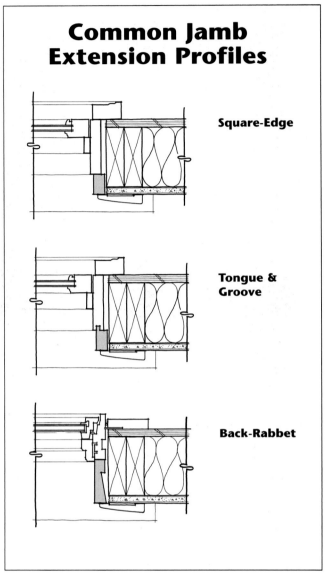

Figure 7-11. *Jamb extension profiles differ, depending on the shape of the window jamb and the thickness of the wall. A square-edge works well on a plain window jamb when the extension is narrow, while some extensions must be milled with a tongue to fit a corresponding groove in the jamb. For thick walls, a back-rabbet reduces the length of the screw needed to attach the extension.*

you use, if the lines are scribed accurately and the cut is made carefully, the piece will fit perfectly the first time *(Figure 7-10)*.

CASING WINDOWS

Installing window casing is often more difficult than working with door casing because window jambs aren't always flush with the finished wall. Window extension jambs are a fact of life for trim carpenters. Although custom jamb widths can be ordered from most manufacturers, sometimes the wall thickness on a new construction site hasn't been determined when the windows are ordered. In a remodel, the wall may have gained thickness — from the addition of rigid foam insulation, for example, or furring and paneling. Some jobs are so complicated that it's difficult to determine all the various wall conditions. But often, we just plain forget to order wider window jambs. And forgetfulness can be costly, because most window companies, including Marvin, Maestro, Eagle, Pozzi,

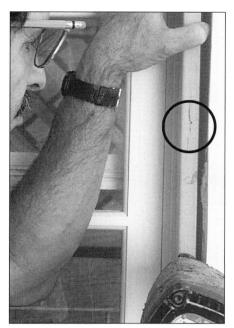

Figure 7-12. MDF is not a good material for making jamb extensions because jamb extensions must be nailed though the edge, and MDF tends to split when edge-nailed, as shown in circle.

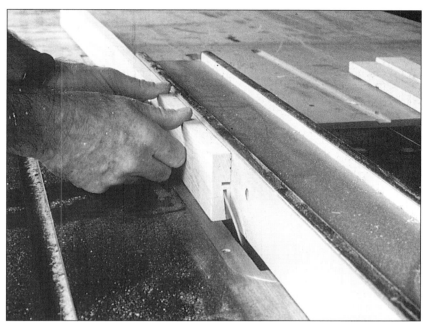

Figure 7-13. The angled finish-cut for a back rabbet on a wide extension jamb is cut on the table saw with the blade exposed.

and Weather Shield, manufacture window jambs with either custom-sized jambs or factory-installed extensions at little or no additional cost.

Window manufacturers handle jamb extensions in one of three ways *(Figure 7-12)*. The first and simplest is a piece of square-cut stock, such as that supplied for a Pozzi window. The second type of extension, used by companies such as Eagle, attaches by means of a back-rabbet in the jamb. Andersen provides a third configuration, a tongue that mates neatly into a corresponding dado in the jamb.

Site-Built Extensions

Whenever I have to mill my own extensions, whether from job-site complications or just plain forgetfulness, I use S4S stock. For paint-grade windows, I like finger jointed pine. While it's tempting to use MDF because it's readily available, mills easily, and paints beautifully, MDF is the wrong material for jamb extensions. MDF is laminated and does have a grain direction. Fastening MDF through the face works well, but nails or screws driven into the edge cause flaking and splitting. And jamb extensions are all about fastening through the edge *(Figure 7-13)*. Extensions are fastened to the jamb through the edge, and then the casing is fastened to the extension by nailing into the same edge. So it's best to use real wood. If the job is being stained, then pick material that matches the grain pattern of the window jambs.

Milling Jamb Extensions

For reasons of ease and speed, I prefer to mill extensions in my little shop, though I often cut them on site, too. Either way, to increase productivity and save installation time, I measure all window and door extensions before I'm called to install the finish work; this way, all the material is on the job the first day of work. After measuring the extensions, I purchase stock that results in the least amount of waste, though I try to avoid ripping more than two extensions from one piece of stock. Using factory-finished surfaces reduces milling time and the need for a surface planer. A table saw equipped with a fine-tooth carbide blade makes a smooth enough cut for the inside edge, which is butted against the window jamb. But sometimes I must rip multiple narrow extensions. I pass these through my portable surface planer before easing the edges with a table-mounted router or hand-held laminate trimmer with a $3/16$-in. roundover bit. If you don't own all those tools, buy the smallest stock you can find and anticipate a little more waste.

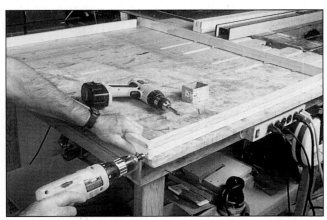

Figure 7-14. *Extension frames that are wider than 1 in. should be preassembled.*

You can quickly mill tongues and back-rabbets with only a table saw. The tongue for an Andersen extension requires several passes, though I often eliminate the tongue for narrow extensions, applying them flat, directly on top of the dado. Back-rabbets for wide extensions eliminate the need for deeply countersunk screw holes. If the extension can be applied before the window is installed, use nails, rather than screws, to fasten the extensions through the shoulder of the back-rabbet. The shoulder cut for a back-rabbet should be made about 1 in. from the edge of the stock. The angled finished cut begins at the outside edge of the stock and meets the shoulder cut *(Figure 7-13)*.

Picture Frame Extensions

The majority of the extensions I apply are less than 1 in. wide, primarily because shear paneling, or sheathing, is often forgotten in a window order. For these narrow extensions, I measure all the windows on the job, pre-cut the pieces to length using a repetitive stop on my chop saw, and then brad-nail them to the jambs. To determine the width of the extensions, I hold a block of wood or a small square flat against the wall and measure from the window jamb to the block or square. Then I add $1/16$ in. I like to mill my extensions so that they're slightly proud of the wall—that extra $1/16$ in. makes it easier to install the casing. I take several measurements on each window, especially at the head and sill, where drywall tends to thicken, then I average the measurements.

But extensions that are wider than 1 in. should be preassembled, with the ends fastened just like the head and legs of a jamb, so that the joints will never spread *(Figure 7-14)*. I approach picture-frame windows, which are cased on all four sides, differently than I do windows with stool and apron (more on installing stool in a minute). Because I can work faster in my shop, with a large waist-high work surface and all the necessary tools within easy reach, I measure extension frames for picture-frame windows while I'm figuring the material take-off. Then making them is a cinch.

Homes today may be large and have numerous windows, but most of the windows are the same size. My frame take-off is usually a short list of window sizes with slash marks for each frame. Occasionally the list will include duplicate window sizes, with different extension widths. The frame sizes are easy to measure, too. I measure the inside dimension of the jamb and add twice for the reveal, about $3/8$ in. for a $3/16$-in. reveal. Adding $1/2$ in. works fine, too, producing a $1/4$-in. reveal. If the window stop is proud of the jamb, then I hook my tape on the stop, measure to the outside of the opposite piece, and add $1/16$ in. to be sure the frame will slip over the stop.

Stacking Jamb Extensions

Doubling up jamb extensions is another method for trimming thick walls. It can be more efficient to use double extension frames, or stacked extensions, rather than wide one-piece extensions, because double frames are easy to preassemble from square-cut stock, and they can be installed in stacks, one after the other, using a nail gun. For double extensions, I use smaller reveals, which allows more wood for fastening. I measure for the first frame just as I do any other window, then add $3/8$ in. for the second layer, which leaves a $3/16$-in. reveal.

I use a reveal block to speed up the installation process. I cut a $3/16$-in.-deep rabbet all the way around a $3/4 \times 3$-in. piece of square wood, so the block can be positioned quickly in each corner, in any direction *(Figure 7-15)*.

I cut and assemble the frames in my shop, or if the second extension is narrow, I attach the pieces one at a time, directly on the window jamb.

Scribing Extensions for Irregular Walls

Framing isn't always plumb, and occasionally a window or doorjamb can't be set parallel or in plane with a wall. Scribing extensions for out-of-plumb or irregular walls isn't unusual. Some trim carpenters install extensions proud of the wall,

Figure 7-15. *A reveal block, rabbeted 3/16 in. on all four edges, speeds the installation of casing (top). Use it to mark pencil lines at each corner that will serve as guides in positioning the inside edge of the casing (above).*

Figure 7-16. *A quick method for scribing a jamb extension is to mark the outside face of the extension with a pencil running flat against the wall.*

then use a portable power plane to cut them almost flush to the wall. If the drywall is installed, a small 1/16-in. spacer taped to the bottom of the plane prevents planing the drywall and ruining the planer blades. But extensions can be scribed and cut to fit, too. I use two methods for scribing extensions before attaching them: One is fast and crude, the other is slow but perfect.

For apartments and tract housing, I position each piece of the frame against the jamb, and lay my pencil against the drywall to scribe a line on the back face of the extension *(Figure 7-16)*. I cut each piece of the frame using my panel saw, then sand the edge (only near the reveal line), and ease it with a roundover bit in my laminate trimmer.

In custom homes I take a little more time, particularly with stain-grade material. Rather than producing edges that follow the irregular wall surface, this method results in straight edges that are either proud of or flush with the wall. First I determine the thickest part of the wall along each side of the opening. I transfer these measurements to the nearest corners and then to the corresponding extension piece end, using the largest measurement at each corner, so that the mating extension pieces are the same width where they join. I draw straight lines on each extension piece between the end marks and cut to those lines using my panel saw or my table saw with the fence removed. The pieces may be tapered, but the edges are straight.

Extensions for Stool-and-Apron Windows

Because window stool is applied directly to the jamb and is both stool and extension, it must be cut to fit before the extension legs and head can be installed *(Figure 7-17)*. Therefore, I approach stool-and-apron

Figure 7-17. *Window stool must be cut and fit before the rest of the extension jam can be installed. Therefore, extensions for stool-and-apron windows are done on-site rather than preassembled in the shop.*

Figure 7-18. *Hold the stool in position and scribe the width of the ears from the wall.*

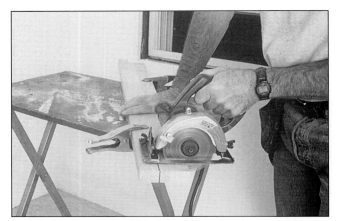

Figure 7-19. *The small blade and high speed of a panel saw help in cutting scribe lines and back-beveling for the stool ears.*

windows differently and never attempt to preassemble them in my shop. Each piece of window stool must be scribed to fit both the jamb and the finished wall, with ears (or horns) that are wide enough and long enough for casing termination. However, production techniques can speed this process, too.

Increasing production is often a matter of reducing the number of tools you must use at one time. Therefore, like every piece of repetitive door and window trim, I first measure and cut all the stool to width and length, then distribute each piece to every window, along with precut extension legs and heads. Next, I scribe and cut each piece of stool. Once the entire stool is cut to fit, I preassemble the stool and extension jambs and then attach the finished frame to the window. This method minimizes the amount of times I change tools, and dramatically speeds an otherwise slow and expensive process.

Cut and scribe the stool. Cutting the stool to length is easy arithmetic. Begin with the inside dimension of the window jamb and add the following: Double the width of the casing, double the casing reveal on the jamb, and double the casing reveal at the end of the stool. Thus, 3 1/2-in. casing, with 1/4-in. reveals at both the jamb and stool, requires a stool that is 8 in. longer than the inside jamb.

Scribing the stool to fit isn't too difficult, either. I begin by using the same casing and reveal measurements. After cutting the stool, I measure in and mark on each end the width of the casing plus the jamb and stool reveals (4 in. for 3 1/2-in. casing).

For the stool ears to fit tightly against the wall and the window jamb, they must be scribed and notched an equal distance from the window jamb.

I hold the stool in place and take a quick measurement to verify that the stool and jamb are parallel. If they're not, I slip a small shim behind one end to correct the problem. Then I spread my scribes the distance between the stool and the jamb, and scribe the ears from the finished wall *(Figure 7-18)*. This scribing technique works well for bullnose walls, too, and always results in a tight fit.

As with other scribe cuts, I use a small panel saw because the combination of high rpms and the small blade helps in following irregular lines *(Figure 7-19)*. I keep the blade square to the stock at the start of the cut, so that the end of the stool will meet the wall square. But I tip the blade slightly and back-bevel the inner portion of the notch, so that the top (visible) face of the piece will fit tightly against the wall without any struggle. I finish the inside corner of the cut with a back saw, also back-beveling slightly.

Install the stool and extensions. Wide window extensions should always be firmly attached to

Figure 7-20. *The floor is a good place to assemble the stool and extensions for large windows. Use the wall as a brace.*

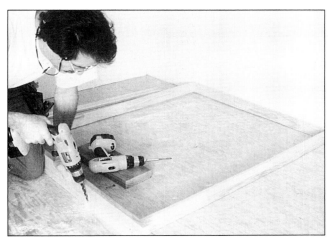

Figure 7-21. *Using a ³⁄₈-in. bit, drill pocket holes in the bottom face of the stool, and finish with a ¹⁄₈-in. bit, drilling pilot holes though the back edge.*

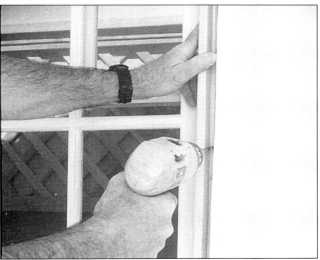

Figure 7-22. *Position the extension frame against the jamb (top) and fasten it with drywall screws through the back bevel in the legs and head (middle), and through the pocket holes in the stool (bottom).*

each other and to the stool. Only then should they be installed on the jamb. The joint between the extension jamb and the stool is especially weak and can spread if it isn't properly secured. On large job sites, I like to assemble the frames on a worktable, which I carry into each major room. (I have an old Ryobi chopsaw stand, but a set of saw horses or a Workmate (and a sheet of plywood work well, too.) For large windows and small rooms, the floor makes a good worktable, with one wall serving as a brace *(Figure 7-20)*. I bring along two cordless drills, one with a long Phillips-head driver and the other for drilling pilot holes. I use a tapered drill bit and countersink for pilot holes, and 1¹⁄₂-in. drywall screws to fasten the legs to the head and to the stool.

While the extension frame is still lying flat on the floor or worktable, I use a ⁹⁄₆₄-in. bit to drill pilot

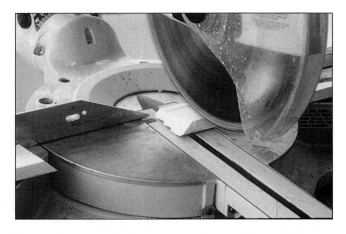

Figure 7-23. *Cutting and attaching the self-return (top and middle) and then installing the completed apron (bottom) is a simple but gratifying finishing touch.*

holes completely through the back-rabbet. The pilot holes are large enough so that the mounting screws just slip in the extension. I angle the holes slightly to make it easier to reach the screw with a driver, but I don't angle the holes too much, or the tip of the screw might penetrate the face of the jamb.

Though the stool isn't back-rabbeted, I still drill a few pilot holes, beginning as pocket holes, from the bottom face of the stool and through the back edge. I start the holes with a 3/8-in. bit, drilling perpendicular to the stool about 2 in. from the back edge of the frame. I slowly raise the drill, bringing the bit almost vertical, and bury the tip in the stool **(Figure 7-21)**. I finish the hole with a long 1/8-in. bit. A pocket hole this size countersinks screw heads so they won't interfere later when it's time to install the apron.

Next, I position the frame against the jamb and use a long Phillips-head driver in a magnetic bit holder to drive 1 1/2-in. drywall screws **(Figure 7-22)**. One screw about every 1 1/2 ft. is more than sufficient. If the drywall is already installed and cut close to the window opening, a well-aimed hammer blow will clear a path to the pilot hole. I also clear the drywall from the pilot holes beneath the stool and screw the stool to the jamb.

Installing the Casing and Apron

Like door casing, the entire window casing should be cut at the same time and installation should begin at the head of the jamb. When picture-framing a window, don't fasten the lower ends of the legs until after the bottom casing is applied, then draw both pieces together so that the miters are tight and the reveals are even.

If the window has a stool, I drive nails up through the stool into the bottom of the casing legs, which provides a little additional support, then I apply the apron. Installing the apron is one of the most gratifying parts of the job. Measure for the length of the apron by pulling a tape across the bottom of the stool — that's the shortest measurement because the front edges of the stool are sloped. The apron should terminate about 1/4 in. short of the stool, so subtract that amount from the length of the stool. Terminate each end of the apron with a matching return, also called a self-return, cut from a short piece of scrap casing **(Figure 7-23)**. That final detail is the easiest and yet one of the most gratifying parts of the whole job.

Figure 7-24. *Use a long level to check reveal lines and keep casing straight across multiple windows.*

Figure 7-25. *Install a filler to straighten out reveal lines.*

CASING MULTIPLE WINDOWS

There are several different ways to case multiple windows. If the mull spaces between the windows (both the vertical and the horizontal spaces between the windows) are wider than twice the width of the casing, then each window can be picture-framed as a separate unit. If the mull spaces are narrower than twice the width of the casing, the mull casings can be ripped down (creating beautiful "trapped" miters), or "mull battens" can be installed between the windows.

Picture-Framing Ganged Windows

No matter which method is chosen for gang-casing windows, it's essential to lay out the perimeters of the casing before beginning installation. Picture-framed ganged windows are attractive if the casing is level across the heads and lines up with adjacent window or door casing. It also looks good if the perimeter casing is parallel to adjacent walls or other casings. And in order to carry the look successfully, the casing must be parallel between each window, too. If any of these criteria are not met — especially if the windows aren't set properly with even and parallel mull margins — then a wall of picture-framed windows can look terrible.

Start determining how a wall of windows will look by running a level line across the heads of the adjoining windows. Pencil the line at the location of the casing reveal on the jamb. If the heads are slightly out of level, then adjust the reveal line a little bit — split the difference from one side of the

Figure 7-26. *Since the stool is thicker than the header, it is easier to fudge a reveal that is different in size across a bank of windows.*

windows to the other side. A slight change in the jamb reveal across a 10-ft. or longer span will be much less noticeable than an abrupt difference in height between two adjacent miters.

I didn't choose the windows in the following photographs intentionally — this stuff just happens — but they turned out to be a great example of the problems a trim carpenter can run into. Using a long level, we quickly discovered that the right flanking window was 1/4 in. higher than the center window frame, and the left flanking window was 1/2 in. lower than the center frame *(Figure 7-24)*.

To solve the problem, I cut a filler for the head of the window jamb on the left *(Figure 7-25)*, and

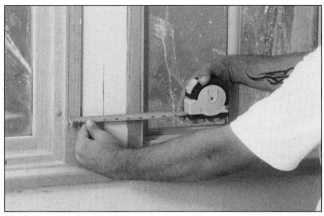

Figure 7-27. *When a pair of mull casings is trapped in a space too small to accommodate their full width, fit them by dividing the mull space in half. Then, use that measurement to rip the casings to width — which also determines how much to cut off the long point of the miter.*

Figure 7-28. *Hold both pieces of casing in one hand and center them in the mull space, watching both reveals as well as the miter joints as you tack them in place.*

lowered the reveal line by 1/2 in. The filler allowed me to cheat the upper casings into a straight line at the cost of a slightly tapered reveal line, which wouldn't be too visible after painting.

The bottom length of stool was easier to lay out, because the thicker sills allowed more room to fudge the reveals *(Figure 7-26)*. The vertical perimeter legs had to be shifted slightly in order to maintain an even margin between the outside of the casing and the adjacent walls.

I cheated the reveals on the center vertical casings, too. Here the trim had to be installed parallel. By following this layout routine, the pencil lines provided a safe and foolproof method for achieving parallel margins, without having to recut or reinstall a single piece of casing (even on this crudely installed series of windows!).

Cutting Trapped Miters

Often picture-frame windows that are ganged are set too close together and there isn't enough room in the mull spaces for full-sized, or even partial-sized pieces of casing between the windows (commonly referred to as back-to-back casing). For picture-framed windows, this situation always results in trapped miters — miters that are cut short because they abut adjacent miters.

In order to maintain both proper molding details and correct miters, an equal amount of material must be removed from each piece of a pair of mull casings. But with carefully prepared layout lines, the sizes of these pieces are easy to determine.

Measure the mull space between the reveal lines penciled on the jamb. Be sure to take a measurement at both the top and the bottom of the window (just in case the mull spaces are uneven). If the mull spaces are even, then the casing is much easier to cut. If not, try to cheat the reveal lines as I described earlier, to avoid scribing the outside edge of the casing (a time-consuming and sometimes frustrating task).

Next, divide the mull spaces in half. That dimension is exactly the size to which the mull casings must be ripped, either before or after they're mitered. The distance from the reveal line to the center or the mull also determines the length of the miter on the head casing: The long point of the miter must be cut off to abut the adjacent head casing. Cut and miter the casings as you normally would, and then cut off the long point exactly the distance from the reveal line to the center of the mull *(Figure 7-27)*.

Tack the horizontal head casing with only two brad nails, so that it can be adjusted slightly. If possible, hold both vertical pieces of mull casing in one hand, adjust the pieces so that reveals on both jambs are equal, then tack them to the jambs simultaneously *(Figure 7-28)*.

Cutting Mull Battens

Mull spaces aren't always wide enough for two back-to-back pieces of casing, even if they're ripped down — extremely narrow rips of casing look awkward and the process is labor-intensive and expensive. Instead, I frequently case the perimeter of

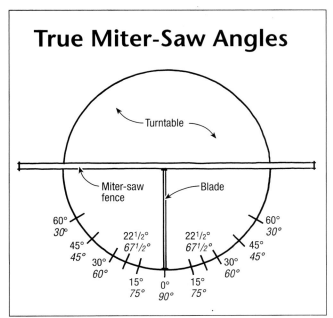

Figure 7-29. *A typical miter-saw scale (upper numbers) does not indicate actual cutting angles (shown here in lower italicized numbers).*

Figure 7-30. *The best way to measure an odd-angled window is with a protractor.*

these units in a picture-frame fashion, and then fill in the mulls with full-width mull battens. I make mull battens from sheets of 1/4-in. MDF. For stain-grade windows, I use veneer core (1/4-in. hardwood veneer with an MDF core). Lengths over 8 ft. can be spliced, though 1/4-in. MDF is available in 10-ft. sheets, too. Occasionally, for top-end jobs, I order sheets of veneer-core in custom sizes — 10 ft. and even 12 ft. long.

If the window is being trimmed with stool and apron, I install a full-length piece of stool first. Then I case the perimeter of the windows as one large picture frame and apply the apron. After laying out the reveals and sizes of the mull battens, I rip the correct widths of stock, ease the edges with a 3/16-in. roundover bit mounted in my laminate trimmer, and brad-nail the pieces in place.

CASING ACUTE ANGLES

Scribing casing is one of those jobs that once made me flinch. But mitering acute angles used to drive me absolutely nuts. It took two sharp finish carpenters — Jed Dixon and Gary Ashburn — to teach me the best trick for mitering acute corners. And with their help in solving this mystery, I finally understood the relationship between my miter-saw scale and my protractor.

Miter-Saw Scales

The scales on power miter saws have confused many carpenters. The problem is this: Set at "0," power miter saws actually cut at 90 degrees. To make matters more confusing, set at "45," miter saws *are* cutting 45 degrees!

Miter saws should come with a warning label: THIS SCALE IS NOT TO SCALE. Because tool manufacturers arbitrarily chose "0" to represent a 90-degree cut, the marks between "0" and "45" have little relationship to actual angles. It seems "0" was chosen as the mark for a 90-degree cut so that miter saws would be easier to use for cutting the common angles to which they are limited, namely "soft" angles that lie roughly between 80 degrees and 180 degrees. Miter saws are not designed for "hard" angles — those that are sharper than 80 degrees — because most saws can swing only far enough to cut miters up to 40 degrees (a few swing as far as 35 degrees). Right now I can hear many readers saying: "What!? My saw can cut to 55 degrees!" Those readers would be right, but for the wrong reasons.

The best path out of this confusion lies in the recognition that miter-saw scales are a white lie: If "0" actually equals 90 degrees, and "45" equals 45 degrees, then "40" is really 50 degrees, "35" is 55 degrees, and so on. I've penciled in these corrections on my miter saw *(Figure 7-29)*. To help straighten my mind out, I try to think of the all-important

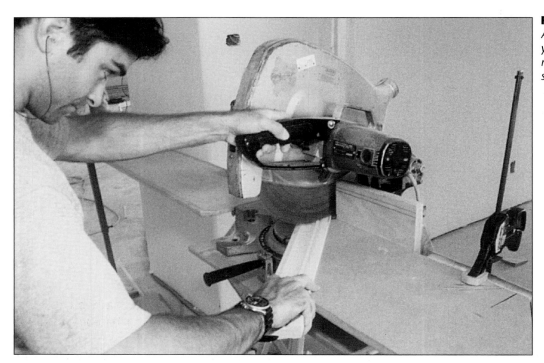

Figure 7-31.
A 45-degree fence allows you to cut a 20-degree miter by positioning the saw at the "25" mark.

Figure 7-32. *It's a good idea to cut test pieces for unusual miters.*

"22½" mark as 67½ degrees, especially when another crew member yells, "Cut that piece at a 22½!"

Finding the Right Angle

Angles can be guessed at and then several test pieces cut, but to save time, especially when the window you're working on is near the top of an extension ladder, it's best to use a protractor. The angle of the window in *Figure 7-30* reads 40 degrees. Split in half, the miters for that angle must be cut at 20 degrees. That's the simple part. The hard part is making a 20-degree cut on a miter saw. You can't just set the piece of molding on the saw, turn the blade to the "20" mark, and get a 20-degree miter — the "20" mark is actually 65 degrees!

An Acute-Angle Fence

I used to cut acute angles by experimenting with different sized wedges between the molding and the chop-saw fence. But that technique was always dicey, full of frightening potential, and never accurate — especially for the opposing miter. Here's the trick I've learned for cutting miters on acute angles.

First, cut a wide scrap of material — the wider the better — at a 45-degree angle. Clamp that board to your miter saw extension wing so that the long point of the miter is up against the fence and just touches the blade. Use the long miter on that board as a temporary fence. With the saw set at "0" the blade will make a perfect 45-degree cut; but with the saw set at "10" the blade will make a 35-degree cut! Simply subtract the "mark" on the miter-saw scale from 45 degrees (the angle of the temporary fence) to arrive at the required miter angle.

The window in Figure 7-31, with a 40-degree angle, required a 20-degree miter. The carpenter simply rotated the saw to the "25" mark. In other words, he subtracted 25 degrees from the 45-degree fence to achieve a 20-degree miter *(Figure 7-31)*. The fit was perfect, the first time. But we still cut test pieces *(Figure 7-32)*.

Chapter 8

Baseboard

Baseboard installation is usually reserved for beginners, and not because it's easy to install. Baseboard offers just the right amount of challenge for a beginner: it's a good job for learning how to visualize and cut miters, how to use a power miter saw, and how to manage a nail gun. Baseboard has only four basic joints, and mastering that joinery is the foundation for most trim carpentry.

Some of this chapter will no doubt bore old hands, but skip carefully — I've collected tips and techniques from almost every carpenter I've met, and even experienced carpenters might learn something new. I believe that new carpenters profit immensely from careful, accurate, and simple instruction; cutting and installing baseboard is the all-important beginning.

BEFORE CUTTING

As I've said earlier, production carpentry isn't a matter of cutting corners or quality, but rather of developing systematic routines that shave time from repetitive tasks. Measuring properly and creating an accurate cut list for baseboard molding are essential routines that can save countless trips to the miter saw, so that energy and effort can be applied toward molding installation, rather than long hikes.

Developing an Efficient Cut List

I carry a small pad of paper in my tool belt so that I can record measurements and odd notes while I'm working. A scrap of wood works almost as well.

Whenever I enter a room that I'm about to base, I always go through the doorway and turn to my right, then continue in that direction around the room. Turning to the right is a habit I developed from many days of coping baseboard (more on coping later).

After turning to the right, I face the wall and measure the first length. I write the measurement in the middle of the line on my note pad. I make a shorthand notation for the right-hand joint to the right of the measurement; I make another shorthand notation for the left-hand joint to the left of the measurement. That way, when I'm at the saw, I don't have to remember each piece of molding and it's easy to determine how to cut each end — the measurement on the right is the right-hand cut and the measurement on the left is the left-hand cut.

I use a large check mark (^) of sorts to identify regular 45-degree miters for typical 90-degree inside corners *(Figure 8-1)*. For butt joints — where baseboard terminates against casing or cabinets — I write a capital "B." I distinguish outside corners

Figure 8-1. *The piece on the left is a butt joint to a 90-degree inside corner, written on the cut list as **B 5½ in.** ^. The piece on the right is a 90-degree inside corner to a butt joint, written on the cut list as ^ **5¼ in. B**.*

Figure 8-2. To terminate baseboard, as at a landing step, use a self-return.

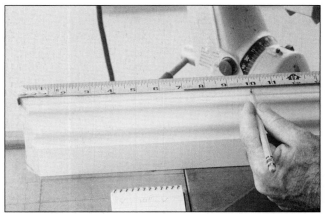

Figure 8-3. Measure an "inside-to-inside" corner with the tape measure hooked over the left-hand long point and numbers right-side up. Inside corners are always measured long point to long point, and the long point is always at the back of the molding.

with a capital "O." If the corner is not 90 degrees, I note the amount of the miter (most often it's "22 1/2," or 67 1/2 degrees, for a 135-degree corner), and I note whether it's an inside corner (^) or an outside corner (O).

In several locations — at the top of a stairway or balcony, or at the edge of a flush jamb or patio door frame — baseboard has to terminate in the field. I cut self-return miters for these terminations *(Figure 8-2)*, and I note them on my pad as "SR."

Measuring Properly

Measuring for a length of baseboard isn't simply a matter of pulling a tape and writing down a measurement. Every measurement requires a little concentrated thought about past experience. Most proficient and accurate trim carpenters have learned that every length of molding must be measured and interpreted. For instance, always cut long lengths long — add 1/8 in. to any length over 12 ft. Measure and cut shorter lengths exactly the right size — read the tape measure to within 1/16 in. and even tighter. And cut extremely short pieces 1/16 in. short, especially the stubby pieces — often less than 1 in. long — that fit between the door casing and an adjacent wall.

CUTTING BASEBOARD EFFICIENTLY

For several reasons, baseboard cannot be cut efficiently by using a chop saw on the ground or with roller stands. Continuous extension wings are a prerequisite for achieving true production efficiency: long, 16-ft. pieces of material need continuous support so that accurate measurements can be taken quickly. Varying lengths must also have continuous support so that time and effort isn't wasted blocking up long pieces or adjusting roller stands, then moving the blocks or the roller stands for short pieces, and so on.

Cutting Inside Corners

Once the material is properly supported, it's time to make a little saw dust. Many pieces of baseboard require a 45-degree miter at both ends. I call these pieces "inside to inside" and always cut the left miter first (so that during the next step I can read the tape measure right-side up and not confuse a "9" for a "6"). I cut the left miter and hook my tape measure over its long point, then measure to the long point of the miter on the right side *(Figure 8-3)*. For all cuts in baseboard (except self-returns), I make the pencil mark at the back of the baseboard, because that's where the measurements are taken on the wall. Beginners should note that inside corners are always cut with the long point at the back of the molding. Next, I swivel the saw and position the blade carefully so that it cuts right to my line.

Positioning the blade on my Hitachi saw is easy. I stand up straight and look down through the window in the guard housing. On my Makita compound sliding saw, I bring the nonmoving blade down to the top of the molding, slide the saw out and move the material until an inside tooth is right on my pencil line, then lift the motor, pull the trigger, and cut the material. (Beginners should start cutting each piece a little

Figure 8-4. *Cutting an outside corner, the pencil mark is still on the back side of the molding.*

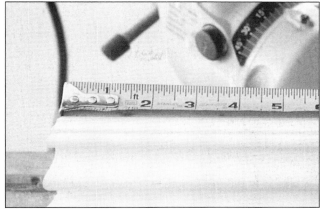

Figure 8-5. *Measuring for an outside-to-outside corner is made easier by aligning the short point of the miter with the edge of the miter-saw extension wing and hooking the tape over the wing.*

long, and then incrementally move the pencil mark up to the blade.)

Cutting Single Outside Corners

Measuring and cutting an outside corner is only slightly harder, but with a little practice, speed and accuracy are easy to achieve. For pieces that have an outside corner on one end and an inside corner on the opposite end, I always cut the inside corner first: it's a lot easier and faster to measure from an inside corner (by hooking the tape over the long point), than it is to measure from an outside corner (you can't hook your tape over a short point, but more on that in a moment). If the outside corner is on the left side, I cut the right side first, even though I have to read the tape measure upside down when I measure to the short point.

Outside corners are exactly the reverse of inside corners: the back of the molding is the short point of the miter, not the long point, but the measurement is still always marked on the back side — except for self-returns *(Figure 8-4)*.

Cutting Double Outside Corners

The next challenge is cutting a piece that has an outside corner on both ends. I used to struggle with these pieces, especially when they were long. Years ago, before I had proper extensions on my miter saw, I used to "burn an inch": I'd pull my tape out, lock it down so it wouldn't retract, then line up the 1-in. mark with the short point of the molding. When I marked the measurement at the opposite

Figure 8-6. *Self-returns are measured from the long point, even though that's at the front rather than the back face of the molding.*

end, I'd add an inch for the one I "burned" at the beginning — if I remembered to. The trouble was, I often forgot. And besides, extending the tape measure and holding both it and the piece of material completely still while marking an accurate measurement just wasn't efficient.

Cutting baseboard is a job where continuous extension wings are truly a delight. I start by cutting an outside corner miter on one end of the piece, and then lay the piece down on the saw table. I align the short point of the miter with the edge of my extension wing, then hook my tape over the edge of the wing, pull the measurement, and mark the material *(Figure 8-5)*.

All this attention to cutting details may seem superfluous, but paying attention to work habits is the best way to increase productivity while simultaneously decreasing mental and physical effort.

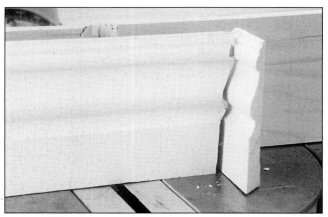
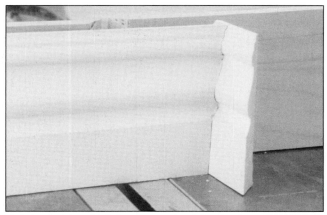

Figure 8-7. *A cope joint, shown both separated slightly (left) and tightly together (right), will remain tight despite seasonal dimensional changes.*

> ## Coping vs. Mitering
>
>
>
> *A glued miter joint in MDF is almost as strong as the material itself.*
>
> This may sand across the grain of some carpenters, but I don't cope baseboard molding nearly as much as I once did. The reason isn't because I'm lazy but because I install more MDF than solid-wood baseboard, and I don't always cope MDF baseboard or crown. I tried coping MDF moldings, but I quickly learned that sharply coped MDF edges crumble when they're compressed. The MDF moldings that I encounter today are also large, at least 6 1/2 in. tall, and they always have elaborate coves, beads, quirks, and other details. Coping these monster moldings is not in my budget, not in the general contractor's budget, and not in the client's budget. Besides, I've learned that tightly fit and glued MDF miters are impossible to break apart without breaking apart the molding itself *(see photograph)*.

Cutting Self-Returns

All measurements for baseboard — whether for 90-degree or 45-degree inside or outside corners — are made and marked on the back of the molding, because the back of the molding touches the wall. But self-returns must be marked on the face of the molding because it's the long point of the miter that must terminate at the proper position *(Figure 8-6)*.

If a piece needs a self-return miter, I always cut that end first (it looks just like an outside corner). Then I hook my tape over the longest point of that long point (the bottom of the molding is usually the thickest, and therefore the longest, point of the miter), and pull my tape across the material to mark the opposite end for its cut.

COPED JOINERY

Like most carpenters, I've always enjoyed the irony of the word "coping." Any carpenter who has mastered the joint recognizes that a good coping technique is often the sign of someone who has learned patience, not just with joinery but also with all the other knots of the business. Coped joints are like padlocks. They are without question the best way to achieve tight-fitting corners. And coped joints ensure that seasonal movement will not cause inside corner moldings to separate (see "Coping vs. Mitering").

Measuring

The easiest and most accurate way to cut a coped joint is to first cut a miter for an inside corner, so I measure for coped joinery almost the same way I do for mitered baseboard. The measurements themselves are nearly identical; I change only my

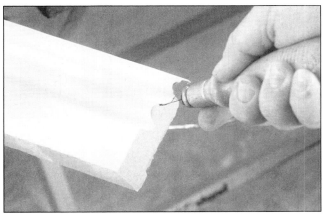

Figure 8-8. *Begin coping at the most fragile part of the profile, usually the top.*

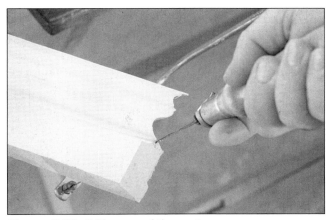

Figure 8-10. *With some of the waste gone, you can resume cutting the profile.*

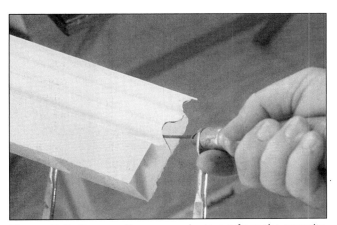

Figure 8-9. *Some sections are easier to cut from the opposite direction.*

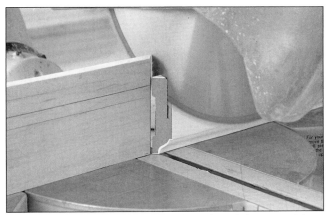

Figure 8-11. *For coping tall baseboards with large straight sections, you can cut most of the waste with a miter saw.*

notations for inside corners. Rather than using a "^," I write "B" for the right end of the first piece, which is installed into the corner first. Then I write "C" for the adjoining cope cut, which is mitered exactly the same as an inside corner, then coped to the right profile *(Figure 8-7)*.

Cutting a Cope Joint

I always enter a room, turn to my right and face the wall, then move toward my left when I'm installing baseboard. This way, the first piece is cut square on the left end, while the second piece is coped on the right end. I'm right-handed and for me cutting copes on the right end is a little easier than on the left end.

Coping is also a lot easier and faster if the material is clamped to a workbench. That's one additional use for the extension wings on my chop saw.

I start the saw blade from the top of the molding because that detail is always the most fragile and often requires a severe back-cut. I hold the blade at a steep angle, and follow the profile of the molding, which is revealed clearly by the miter *(Figure 8-8)*. I maintain a fairly fast stroke, but apply only slight forward pressure on the blade, which allows the saw to turn a tight radius. For some portions of the profile it's easier to back the blade out of the kerf and cut back in the opposite direction, starting from the back of the molding *(Figure 8-9)*. With the waste gone, you can resume cutting along the profile line in the original direction *(Figure 8-10)*.

It doesn't take long to cut copes in soft woods and paint-grade material, such as poplar. But coping large pieces of hardwood and MDF can be a miserable job. One carpenter taught me a nice trick for coping tall MDF baseboard. After cutting all the 45-degree cuts for one room, turn the pieces over and cut through the straight portion of the baseboard with the miter saw *(Figure 8-11)*. Then cope the profile with a coping saw. Another tool that simplifies coping large baseboard molding is a jigsaw

BASEBOARD 119

The Collins Coping Foot

You have to keep your eyes and ears open in this business — someone is always coming up with a better material, an improved technique, or a vastly superior tool. The Collins Coping Foot® is one of those tools.

Manufactured by The Collins Tool Company (Appendix 1), the stainless-steel coping foot mounts quickly on my Bosch jigsaw, and fits many other professional tools as well. Equipped with this custom base and a proper blade, coping becomes an almost effortless task, no matter what type or size of material.

Following the manufacturer's recommendations, I use a Bosch T244D blade, which has six aggressive teeth. But since I do all my cutting from the back of the molding, the front edge is always sharp and clean.

Learning to use the tool requires some practice. I've had the advantage of a lesson from Dave Collins, the fellow who invented and manufactures the coping foot, and that lesson helped shorten the learning curve. Using the coping foot changes the physics of a jigsaw, especially since the coping foot works best with the jigsaw upside down.

The trick to this tool accessory is in the approach: don't attempt to cut the entire cope from only one or two directions. Instead, start by making several perpendicular cuts into the material along each step in the molding, at the top of the molding, and into the deepest section of the S-curve *(Figure A)*.

Next, switch directions and cut from top down toward the lowest perpendicular cut *(Figure B)*. Remove the waste, and then cut from that position up towards the next relief cut *(Figure C)*. Repeat that process until the profile is cut completely, and then finish by cutting the straight section.

Figure A. *When using the Collins Coping Foot, start by making relief cuts at each step in the molding profile and into the deepest portion of the S-curve.*

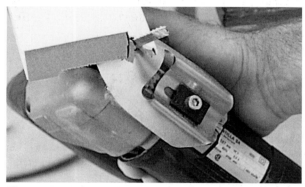

Figure B. *Next, cut from near the top down towards the first relief cut.*

Figure C. *After removing the waste, cut up towards the second relief cut.*

equipped with the Collins Coping Foot®. I use the coping foot whenever I join large stain-grade moldings (see "The Collins Coping Foot").

INSTALLING BASEBOARD EFFICIENTLY

To save needless and repetitive trips to the saw, I cut as many pieces of baseboard for a single room as I possibly can before I begin installing the material.

Once I'm on my knees with a nail gun, I like to stay down and continue all the way around a room.

After placing each piece near its location — assuring myself that I've cut every necessary piece — I start by installing the smallest pieces in the room first, especially those that are trapped between the door casing and a nearby wall. I don't like to cut or install very small pieces with coped ends unless it's necessary. Instead, I've learned to achieve tighter joints and faster production by coping both ends of the previous piece, which means I usually install all the small pieces first. Even if the inside corners are mitered, I install all the small pieces first, so that I don't have to struggle with sliding and working them down from above.

After the small pieces are in, I start at the beginning of my list and fasten the base to the walls. I angle the lower nails down, so they penetrate the sill plate, and then I tap on the wall or use a magnet or even a stud finder to locate good backing for the upper nails. I like to secure the base, top and bottom, at least every couple of feet.

Inside mitered corners usually close up tight, but if they don't, I carry caulk for paint-grade molding. I cope hardwood baseboard, and those joints are always tight. I glue, carefully sand, and, if necessary, fill all mitered splices. Finally, I gently ease the edges on every outside corner (see "Round Corners," next page).

BUILT-UP BASEBOARD

At one time, baseboard molding was nearly always built up from two pieces of material. A broad flat board was installed first, which represented the flat base of a classic column. A small piece of molding or cap was added to trim the top of the lower board, representing the scotia and torus.

Most baseboard moldings no longer need to be built up from a combination of simpler moldings because a multitude of profiles and sizes are readily available in a variety of materials. But realism, historical integrity, and current trends in combining architectural styles require the continued use of built-up moldings.

On a recent job we combined three separate moldings to form a tall neo-classical baseboard. This installation exemplifies the techniques of stacking multiple moldings into a single baseboard profile. In the paint-grade sections of the home, we used 1/2-in. MDF as the flat board, a rabbeted poplar panel molding for the center cap, and a decorative bead-and-barrel design for the final trim. In the stain-grade sections of the home, we ripped 8-in.-wide strips of 1/2-in. mahogany plywood (with an MDF core). We ordered custom 10-ft. and 12-ft. sheets of this "plywood" from Forest Plywood, a full-service national supplier of plywood and laminate sheet goods (Appendix 1). Major plywood suppliers will lay up custom orders in many sizes and grain patterns. Having custom sheets of mahogany laid up was the best way for us to ensure that grain and color matched throughout the home. We also had to

Figure 8-12. *This three-piece built-up baseboard begins with butt-joined hardwood plywood (top). The plywood edge is capped by a rabbeted panel mold (middle). A bead-and-barrel molding is installed last (bottom).*

order panel molding, custom-milled in mahogany, to match the stock poplar design.

Of course, the mahogany MDF-core was installed first. The inside corners didn't have to be coped or mitered, just glued and cut square for butt joints. We installed the rabbeted panel molding next, mitering and gluing the corners. Finally, we installed the decorative bead-and-barrel molding on top of the panel mold *(Figures 8-12)*.

Round Corners

Not many years ago, I cut three pieces of baseboard for all round outside corners. The joints between the pieces *(Figure A)* were cut at 22¹/₂ — a 67¹/₂-degree angle! But bullnose corner beading has since become ubiquitous and round baseboard corners are now made in a variety of materials, especially wood and resin, and are stocked to match nearly every molding pattern.

To measure baseboard with round corners, I make a few slight changes in my cut list routine. Before starting a cut list, I trace lines on each wall using a round corner piece *(Figure B)*, and then I measure and make my list. I mark those corners as "B" for butt or straight cuts.

Even though round corners are manufactured to match stock moldings, the profiles rarely match perfectly. So, rather than installing all the corners first, I install each corner as I get to it, along with both pieces of adjacent baseboard. That way, I'm better able to align all three pieces so that the joint is smooth and seamless.

I apply adhesive caulk on the back of the corner, press the corner lightly against the wall, then glue the end of one piece of baseboard and hold it against the wall and the corner. Usually the corner has to be pressed tighter against the wall in order to align the profiles. Once I have the pieces flush, I fire a nail or two through the baseboard — just to hold it still, but I don't nail the corner yet. First I align the second piece so that all three pieces fit around the corner with the least amount of sanding. Then I nail everything to the wall.

I prefer to nail the corner itself with 1-in. brads. Longer nails never do a bit of good because the backing is too far from the drywall bead. And longer nails have a habit of glancing off the round corner beading and then sticking out of the wall and interfering with one of the other two pieces of baseboard (or interfering with my finger if it's too close to the nail gun!). A few brads are great for saving fingers and securing the corner block until the adhesive caulk cures.

Figure A. *One way to cut baseboard for a bullnose corner is in three sections. Each is mitered to "22¹/₂" degrees, glued, and carefully nailed.*

Figure B. *If you're using round-corner baseboard, mark the position of the round section first (top), and then measure and cut the abutting baseboards to those marks. Apply adhesive to the back of the corner (middle), and put it in place. Position all three pieces for a seamless fit before nailing them in place (bottom).*

Figure 8-13. *Laminating stain-grade radius baseboard on-site requires an array of cleats and braces.*

RADIUS BASEBOARD

Bullnose corners are everywhere today, and full-radius walls are popular, too. New products have made trimming some tricky walls easier, but occasionally only old techniques seem to work. I use two methods for installing baseboard on radius walls, depending on whether the trim is paint-grade or stain-grade.

Paint-Grade Radius Baseboard

Fortunately, most of the trim that's applied to radius walls is paint-grade, which can be easy to install. I use flexible trim for paint-grade applications. My local supplier stocks flexible moldings in a wide assortment of standard baseboard profiles. No special ordering information is required for flex-trim baseboard either, because it's sold as "straights" in any given length. Flexible baseboard can be shaped to any radius, no matter how small, especially if the material is warm.

Before installing any flexible molding, I lay the material on a flat surface in the sunlight to warm the resin. Warm resin is easier to shape around curved walls, and warm resin won't bruise, craze, or crack as easily from air-fired nails.

Rather than measuring these special pieces, I find it more accurate to mark them in place. I miter one end (if that's the required cut), then hold it tightly in the corner while spreading the remainder of the piece around the wall. Outside or convex walls are the easiest. I walk my hands around the wall, holding the base firmly to the wall, then mark the far end for the proper length, and cut. If the wall is irregular or material is stubborn, I fire a nail wherever necessary to temporarily secure the molding to the wall.

Inside or concave walls require a little more effort — one pair of hands is never enough. Stickers can be cut and braced against nearby walls, or braced off a cleat nailed to the floor. But I usually opt for a faster installation technique: I just nail the material to the wall. Of course, I use as few nails as possible (to save the painter some work), but I use enough nails to hold the molding firmly in place so that I can mark the opposite end accurately.

Stain-Grade Radius Baseboard

Running stain-grade baseboard on radius walls is not easy, and it can't be accomplished quickly. For elaborate molding profiles, I order laminated radius baseboard from a custom millwork shop. I specify the exact radius or diameter, along with the length, and in most cases I also supply a paper or 1/8-in. lauan template.

However, for simple molding profiles and especially flat baseboard, I laminate the material right on the job site. I don't have the proper setup to re-saw 1/4-in.-thick strips of hardwood, especially when they're over 4 in. tall, so I order the material from a local millwork shop.

In most cases, I try to make the radius pieces before the finished flooring is installed, so that I can temporarily screw bracing cleats to the wood subfloor or concrete slab. But time isn't always on my side. On the job pictured in the following paragraphs, I had to make up the baseboard after the

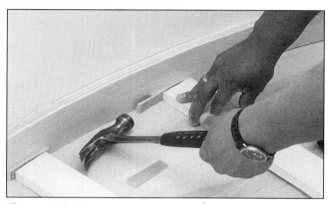

Figure 8-14. *Leave approximately a 1/2-in. space between brace and baseboard for driving shims to apply pressure.*

hardwood floor had been installed, but fortunately before it was sanded and finished.

I always start the installation by gluing up the three strips (for extremely small diameters I use four 3/16-in. strips). I spread Titebond II glue — on both faces of each piece — using a foam "weenie" roller. To hold the pieces together temporarily and make it easier to work them into the corner, I shoot one or two 1-in. brads right at the middle of the laminated length. I use A-clamps to secure both ends so that the boards can slide past each other as they're forced into the shape of the radius.

Laminating radius baseboard always requires more braces than I think it will, so I secure several cleats to the floor, parallel to the wall and about 2 ft. or 3 ft. back — each cleat can support several braces *(Figure 8-13)*. It isn't necessary to fasten the cleats with more than two or three small finish nails, because the pressure of the braces is directed laterally at the nails. If the braces require extreme pressure, use thinner stock and more laminations.

I cut each brace individually, guessing on the best angle from the cleat to the wall. Each brace should be about 1/2 in. short of the distance between the cleat and the baseboard to allow room for a shim between the brace and the molding *(Figure 8-14)*.

Start forcing the material tightly to the wall at the center of the radius, and then tap the top with a block of wood and a hammer, to align the pieces and flush up the top of the stack. Before moving on to the next brace and shims, shoot a brad nail long enough to penetrate the three pieces, which will stop them from slipping vertically. Next, position braces on either side of the center brace, close enough to force the molding evenly and tightly against the wall. After shimming the braces, tap the top of the molding again to ensure that it's flush, and fire another brad near each brace. Continue in the same fashion until the piece is secured to the wall.

I like to leave the braces in place for a day or two to be certain that the glue dries thoroughly. Before removing the braces and the molding, I mark the ends for their appropriate cuts.

The top of the molding is never perfectly flat, as each board slides up or down a little as it bends around the radius. Smoothing off the top of the molding with a portable plane is easy, though holding the material requires several hands *(Figure 8-15)*. The design of this home called for a square-cut molding, though after the molding has been planed smooth, it's easy to add a simple profile with a bearing-guided router bit.

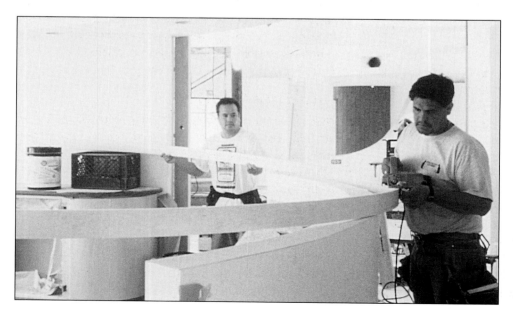

Figure 8-15. *When the glue has dried, plane the top of the baseboard.*

CHAPTER 9

Closet Shelving

Trimming closets used to be simple: You just installed a single shelf and pole in every closet, about 66 in. from the floor so a dress wouldn't drag on the carpet. Maybe people didn't have so many clothes then, but today, closet design is an important part of construction. With a variety of popular shelving configurations, finishing a closet can be a challenge. However, as with all aspects of finish carpentry, careful layout is key. And while efficient use of storage space remains the top priority, efficient shelving installation is a must for a productive and profitable job.

CLOSET SHAPES AND SIZES

Though closets seem to come in many different sizes and shapes, they're actually either one of only two basic types: walk-in closets and reach-in closets.

Walk-In Closets

Walk-in closets come in several shapes. Large walk-in closets allow for three complete walls of shelving and are often referred to as U-shaped. Because they commonly share dividing walls with other closets or bathrooms, most walk-in closets are at least 5 ft. to 6 ft. deep, which allows for two—and sometimes three—types of shelving on each side wall. Additional shelves installed on the rear wall complete the U-shape. Unfortunately, the shelves on the rear wall must be 24 in. short of each side wall, or they'll interfere with the clothes on the side walls *(Figure 9-1)*. This means that in a 5-ft.-wide walk-in, a bank of shelves on the rear wall can be only about 12 in. long.

Step-in closets are abbreviated versions of walk-in closets, with only enough room to step in and back out. Planned effectively, however, step-in closets can still accommodate bundles of clothes. Because the side walls are often less than 4 ft. deep, only one type of shelving should be installed on each side wall (vertical dividers or supports are unnecessary if the shelving span is less than 44 in.). But there's still room for a bank of shelves on the back wall.

Though architects avoid them today, older walk-in closets come in odd shapes, too. Occasionally a walk-in closet is so narrow that there isn't room for clothes on both side walls. These are best configured as L-shaped closets. Always devote the longest wall in an L-shaped closet to the main closet pole. Then begin any shelving on the rear wall at least 24 in. from the side wall with the pole, to allow room for hangers and clothes. There may be enough space for only 18 in. of shelving, but remember, five 19-in.-long shelves provide almost 8 ft. of shelving.

Reach-In Closets

Reach-in closets have only one wall of shelving, which you access while standing in the closet doorway *(Figure 9-2)*. Built in many sizes, reach-in closets are usually only slightly wider than standard sliding door or bi-fold door openings—4 ft., 6 ft., and 8 ft. In a typical 9-ft. reach-in closet, a multiplicity of storage fixtures can be designed, including single poles, double poles, sweater shelves, shoe shelves, and even a few drawers. The user's needs should be considered carefully.

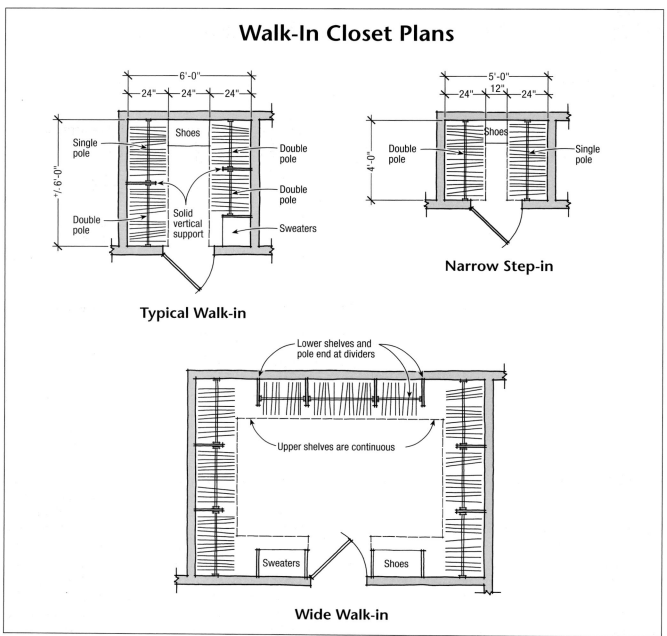

Figure 9-1. *A 6-foot-wide walk-in (top, left) provides room to enter and access 2 feet of shelving along the back wall. A 5-foot-wide closet (top, right) can accommodate a short shelf along the back wall, but does not provide true walk-in space.*

SHELVING SHAPES AND SIZES

Like typical closet shapes, closet shelving breaks down into only a few basic types: single shelf-and-pole, double pole, sweater shelves, and shoe shelves *(Figure 9-3)*. These days, drawer banks are often installed in closets, too. Several of these features can be found in a single closet.

Single shelf-and-pole is a standard. To accommodate long coats and dresses, a section of single shelf-and-pole should be installed in every closet (closets for children often are an exception). To keep dresses and coats from dragging on the floor, install a single shelf-and-pole at least 66 in. from the floor — take the measurement from the bottom of the shelf, which puts the pole at about 64 in. from the floor. For exceptionally tall people, increase the height to keep long clothes off the floor.

Double poles have become more useful. Today, men aren't the only ones in the family who wear the pants. This change in dressing habits has precipitated a change in closet design, too. If pants are folded over a hanger, they need only half the hang-

Figure 9-2. *Reach-in closets are slightly wider than standard sliding-door or bifold openings — typically 4, 6 or 8 feet. A mixture of single and double poles, shelving, and drawers can be used to fit the storage needs of the owner.*

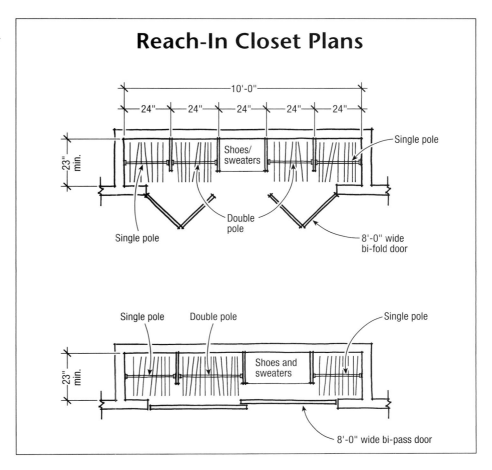

ing height of a long dress — about 34 in. from the top of the cleat to the floor. (Though I rarely install a shelf above a lower pole, I always install cleats, just in case a client wants a shelf later. The cleat also provides a better method for securing a midspan pole support.) Shirts require 40 in. from the bottom of the shelf. Because most of the clothes in today's closets are pants and shirts, double shelf-and-pole at a 40-in. spacing should predominate in every closet, which doubles the storage space.

To make the job of installing shelves easier and to allow homeowners the choice of changing the arrangement of their clothes, I separate all double poles by 42 in., which makes the top shelf 84 in. from the floor. The top shelf should run completely across the closet and around all three walls in a U-shaped closet, so the same 84-in. height determines the second or top shelf over a single shelf-and-pole, too. In most 8-ft. closets, 12 in. of space remains between the top shelf and the ceiling, which is enough room for shoeboxes, hatboxes, and other storage.

Sweater shelves are a must. Sweater shelves are an indispensable type of shelving for every closet, and they're handy for storing more than just sweaters. If dresser space is limited, sweater shelves can double as drawer space and easily accommodate sweat pants and shirts, as well as odd-shaped or bulky items like winter hats and gloves. A typical bank of sweater shelves should begin 16 in. from the floor, which allows room for tall boots on the floor. Succeeding shelves should be spaced about 12 in. apart. If the top shelf is installed at 84 in. from the floor, this sweater-shelf arrangement should result in a somewhat even spacing.

The only shelf in a closet that won't align horizontally with other shelves is the single shelf-and-pole, because it's set at 68 in. from the floor. The 16-in. space between the single shelf-and-pole and the top shelf can be divided again by an additional shelf, which creates a perfect location for a few pairs of shoes.

Don't forget all those shoes. Shoes require only about 7 in. of height (that includes high-tops and pumps). To get the most from your closet space, design shelving specifically for shoes and don't rely on standard 12-in. spacing. An 84-in.-tall bank of shelves, with the first shelf 16 in. from the floor, can include four shoe shelves and three sweater shelves *(Figure 9-4)*. Or, if there's room and shoes

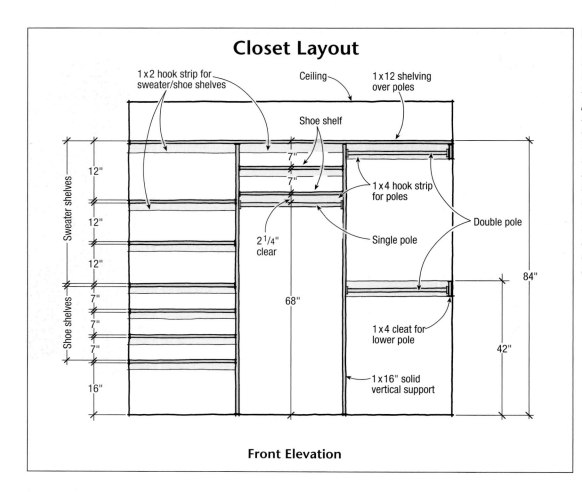

Figure 9-3. *Closet storage breaks down into a few basic types: single shelf-and-pole, double pole, sweater shelves, and shoe shelves. Double poles are more useful than single for modern lifestyles. I typically run shelving around the entire closet at 84 inches, allowing about 40 inches of clear space between double poles.*

enough, build an entire bank of shoe shelves. Of course, flat shoe shelves can be used to store other things, too, like shirts.

Shoe shelves can also be built solely for shoes and sloped so that the complete toe of each shoe is visible at a glance. Some designs call for a small piece of base shoe (also known as shoe molding) installed at the back of the shelf, to create a stop for each shoe heel to hook over. Other designs require a 1-in.-tall edge band, which also stops the shoes from falling off the shelf.

PAINT-GRADE AND PRE-FINISHED SHELVING

For many years the only type of shelving I installed was #2 pine. I often wish I were still using pine because of its strength and ability to take nails and screws on edge without splitting or peeling (the way manufactured panels do). However, lumber prices have almost ended the use of pine shelving, except for stain-grade purposes. In the place of pine, there are several types of shelving products marketed today: unfinished MDF, unfinished particleboard, and pre-finished shelving.

Unfinished (Paint-Grade) Shelving

I prefer MDF to particleboard for unfinished shelving because MDF requires less preparation and paints out more smoothly than particleboard. MDF is easy to cut (with a carbide blade), and, though it's not friendly to hand-driven nails, it is simple to install with a nail gun or screws. MDF is also stronger than particleboard, strong enough to carry the weight of boxes and clothes if the span is less than 44 in. Generally, I use particleboard shelving only when no other material is available from the supplier or contractor.

Unfinished shelving is easier to install than pre-finished shelving — particularly if you're the carpenter and not the painter. I install unfinished shelving in new construction where the shelving and the walls will be painted after all the trim is installed. Because the walls have been only primed, I can draw clear, visible layout lines without it being a problem *(Figure 9-5)*.

Pre-Finished Shelving

Pre-finished shelving has been growing in popularity. Most lumberyards and warehouse outlets now

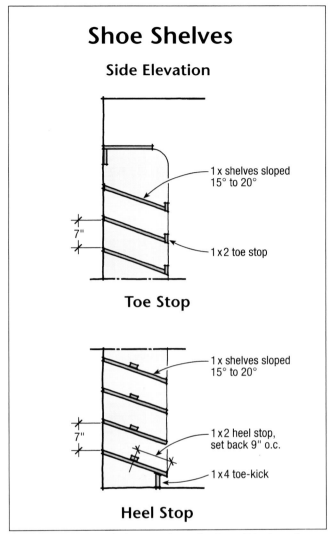

Figure 9-4. *Shoe shelves require only about 7 inches of height. Sloped shoe shelves add a custom touch and make the complete toe of each shoe visible at a glance.*

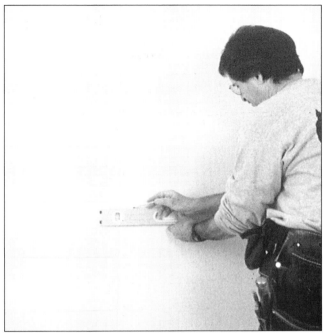

Figure 9-5. *Because MDF shelving will be painted after installation, layout can be done with clearly visible lines.*

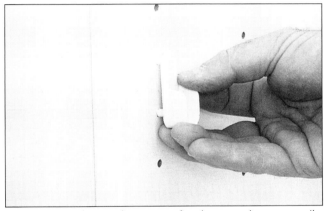

Figure 9-6. *Closet pole supports for chrome poles snap easily into the predrilled holes.*

stock large quantities of pre-finished shelving products, called either kortron or melamine. While both types of shelving often have MDF or particleboard as their core, there is a big difference in the surface finish. Kortron is a white paint that is electronically baked onto the substrate. Melamine is a low-pressure laminate: a vinyl-like paper impregnated with melamine resins and heat-pressed onto the substrate. Melamine is the more durable finish.

SHELVING ACCESSORIES

Pre-finished shelving is especially popular for remodeling closets that are in use because, once the shelving has been installed, the clothes can be replaced without waiting for the paint to dry. And an assortment of shelving sizes and handy accessories are available that make design and installation easier and cleaner.

Most outlets carry melamine shelving in different widths, from 12 in. up to 24 in. Some boards are 16 in. wide and predrilled with holes on 1 1/4-in. centers. These boards can be installed as vertical dividers every 3 ft. to 3 1/2 ft. Horizontal shelving arrangements can then be adjusted or altered easily just by moving a few pegs.

Chrome closet-pole supports are also available *(Figure 9-6)*. These rectangular "rosettes" fit neatly into the predrilled holes in the vertical dividers so that fasteners aren't required and pole locations can be changed at any time. Retail outlets stock many

CLOSET SHELVING

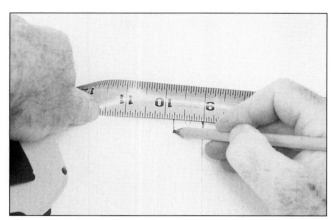

Figure 9-7. *Mark the thickness of each divider so that gang-cut multiple-shelving components will fit perfectly.*

types of fasteners for pre-finished shelving, which I'll talk about momentarily, but first let's consider an efficient method for installing paint-grade shelving, the mainstay for most new construction.

INSTALLING PAINT-GRADE SHELVING

Like all aspects of carpentry, there's a logical sequence to installing shelf-and-pole; many profitless jobs have resulted from overlooking that sensible order. The sequence and speed of installation is also dependent on the design of the shelving components.

Cutting and installing closet shelving is much faster if multiple-shelf areas—like shoe shelves, sweater shelves, and double-pole shelving—are cut to similar even sizes. All multiple-shelf units should be cut in "packages" (gangs of cleats, shelves, and even dividers), so that repetitive stops can be employed as much as possible. To simplify layout and installation, the sizes of these packages should be decided before beginning any layout work. Layout and cutlists are further simplified if components are kept to standard sizes: 24-in.-long shoe shelves, 32-in.-long sweater shelves, 36-in.-long double-pole shelves, etc. Those standard lengths can remain constant throughout an entire project, and they can be pre-cut by apprentice carpenters while a more experienced hand completes all of the layout work.

Layout

Begin by locating each of the vertical dividers. Because I use uniform sizes for multiple-shelf areas, I always begin in a corner with a shoe shelf, sweater shelf, or double pole; I always finish in a

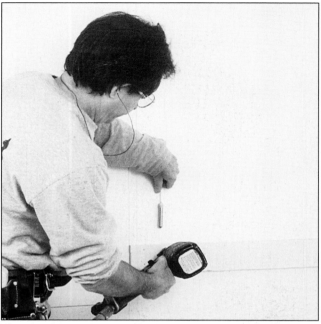

Figure 9-8. *With pre-cut packages of cleats, shelves, and dividers nearby, begin installation in a corner, attaching the cleats for a multiple-shelf section.*

corner with a single shelf because that's the only "special" measurement I'll need to take.

Mark the thickness of each divider *(Figure 9-7)*, so that the next divider can be located accurately and shelving packages will fit properly. After all the dividers are located, use a level to draw plumb lines at each location and place an X on the side of the line where the 3/4-in.-thick divider goes. Then mark the locations for the shelving and use a level to draw horizontal lines for each shelf. The line across the top of the dividers should extend from corner to corner, and in walk-in closets, completely around all the walls. Wherever a closet pole will be installed, mark the wall for a 1x4 cleat, which is wide enough to accept a closet-pole rosette. Use 1x2 cleats for sweater shelves and shoe shelves—any shelf that doesn't have a closet pole beneath it.

Installing Dividers and Shelving Cleats

Be sure to cut all the pieces for every closet before beginning the installation. Measuring and cutting the pieces is easy because only the last shelves and cleats should require measurement in place; the rest of the shelving should be cut in uniform packages.

Always start installation at a multiple-shelf corner. Nail the cleats to the wall first. Use a stud sensor or magnet to locate the studs so that the cleats will be well secured to the wall *(Figure 9-8)*.

Figure 9-9. *Next, fasten the first vertical divider by nailing through the divider into the cleats.*

Figure 9-10. *Follow that divider with the next set of cleats, and continue across the wall until the last set of cleats — which must be measured in place.*

Next, nail the first divider into the ends of the cleats *(Figure 9-9)*. I pre-cut all dividers the same length, so if the floor dips, I shim the bottom until the top of the divider is flush with the top of the cleat. If the floor rises, I cut the divider to fit.

After the first divider is secured, install the next set of cleats and then the following divider, and so on until the last set of cleats, which must be measured in place *(Figure 9-10)*.

Once all the cleats and dividers are secured to the wall, I install cleats also on the dividers, using $1^{1}/_{4}$-in. nails and glue *(Figure 9-11)*. I prefer supporting the shelves with cleats rather than with nails driven through the dividers into the ends of shelves because MDF and particleboard panels easily flake and split when nailed on edge. Cleats support the weight of the shelves better and make for a more durable construction. I cut the cleats with a repetitive stop on my chop saw about $1/4$ in. to $1/2$ in. short of the front of the divider, so the job looks clean and neat. For melamine cleats, I tape the ends.

Installing Shelving and Poles

Once all the dividers and cleats are secured, start installing the shelving. Begin with the shelves nearest to the floor — it's easier to get a nail gun in and properly secure each shelf down to the cleats

Figure 9-11. *For secure shelving, rely on cleats rather than nails — you never know who's going to use those shelves as a ladder.*

CLOSET SHELVING

Figure 9-12. *Always install shelving beginning with the bottom shelf so there's enough room to use a nail gun.*

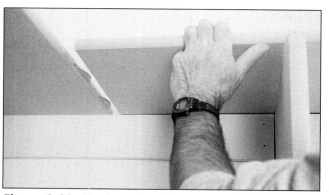

Figure 9-13. *Biscuits are best for supporting corners. Use two biscuits for additional strength (top). Glue and insert the biscuits in the first shelf and set it in position (above). But don't nail it down. If the second shelf is trapped by a wall, you may have to lift the first shelf in order to slip the second shelf over the biscuits.*

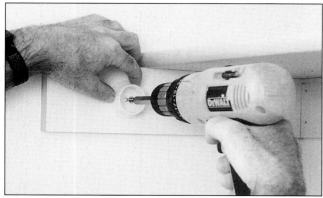

Figure 9-14. *Fasten round rosettes centered 2 1/4 in. below the top of the cleat, positioned so that the back of the pole will be about 1/2 in. in front of the shelf.*

if you start at the bottom and work your way up **(Figure 9-12)**. Install the continuous top shelf last.

Several methods can be used to strengthen the inside corners of the top shelf. H-shaped aluminum support brackets are available that slip between the joint and around each shelf, but the brackets are an eyesore even after they're painted. I've found that biscuits are the easiest method, and they create a strong joint. Before I install the first shelf in a corner shelf, I cut slots for the biscuits **(Figure 9-13)**. I've never had these joints fail, probably because a vertical divider or support isn't usually too far away.

Install the closet poles last. Attach the rosettes so that the center is about 2 1/4 in. below the shelf (or for lower poles, below the top of the cleat), which positions the top of the 1 1/4-in.-diameter pole well below the bottom of the shelf. The back of the rosettes should be 12 in. from the wall — or 1/2 in. in front of the shelf — to allow room for hangers to clear the bottom of the shelf **(Figure 9-14)**.

After the rosettes are installed, measure and cut each pole long enough so that it fits snugly into the rosettes, but not so tight that the pole pushes the dividers out of line.

PRE-FINISHED SHELVING

Pre-finished shelving takes a little more time: heavy lines can't be drawn on the walls because, usually, the interior painting has been completed. Fastening for cleats and dividers must be planned in advance to avoid unsightly marks and blemishes. And all nail holes must be filled, and all screw holes must be capped.

Another reason why pre-finished shelving takes more time to install is that it has a banded edge, so all cuts must be finished with iron-on

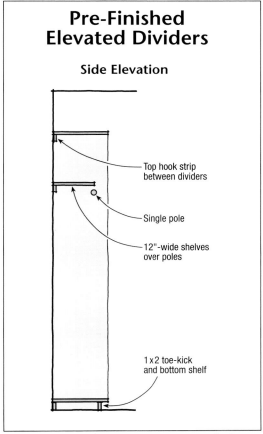

Figure 9-15. *An alternative to tapering the bottom of shelf dividers is to set them on a continuous bottom shelf that rests on a toe kick. This forms a handy storage space for shoes and boxes.*

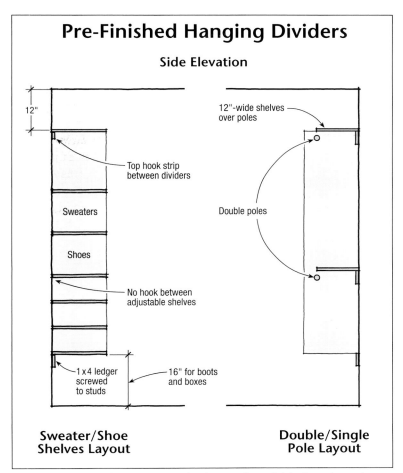

Figure 9-16. *Shelf dividers can also be set about 16 inches above the floor, hung from ledgers. This is often the quickest method.*

tape. I often cut melamine dividers at an angle to the floor, just like I do MDF dividers, so I carry an iron in my van for banding edges on the job (see "Iron-On Edge Tape," page 136).

Of course, the advantage is that once pre-finished shelving is installed, the job is complete and there's no need for paint. Though the material is much more expensive and installation is a little slower, both layout and installation of melamine shelving can be accomplished simultaneously, which amounts to a net savings of time.

Designs for Pre-Finished Dividers

An alternative to tapering the bottom of dividers in a 1-in. footprint is a continuous shoe shelf that rests on a toe kick (see "Cutting Dividers," next page). Such a base shelf provides a perfect foundation for wide dividers, as well as a neat place for shoes and boxes that are otherwise stored on the floor *(Figure 9-15)*. It's also a good point for the carpet to terminate, and the additional shelf diminishes the distance between the top shelf and the ceiling by only the height of the toe kick.

To install a toe-kick and shelf on the floor, first nail a piece of 1x2 right on the floor and against the wall. Use another 1x2 at the front—that's the toe-kick—but place it about $1/2$ in. back from the anticipated front edge of the bottom shelf so there's room for the carpet to tuck beneath the shelf. If the carpet is already installed, it can be cut back and tucked beneath the new shelf, but first remove the old tackless strip and re-use it near the front of the shelf. Once the bottom shelf is secured, the dividers can be installed.

Dividers can also be cut square at the bottom and "hung" from the wall — about 16 in. off the floor, or at the height of the lowest shelf *(Figure 9-16)*. This method is often the quickest and cleanest.

ADJUSTABLE SHELVING

Melamine dividers are available with several styles of predrilled holes on $1 1/4$-in. centers. Some boards

Cutting Dividers

Some contractors and homeowners prefer not to have dividers in the way of the floor covering. I often cut the bottom of a divider at a 45-degree angle to leave only 1 in. of the divider standing on the floor. One inch provides enough support and also allows room for the baseboard to terminate at each divider *(Figure A)*. For melamine dividers, I finish the edge with iron-on tape.

In many installations I also cut back the tops of the dividers. Shelving over closet poles is often 12 in. wide, so that it won't project too far over the pole and interfere with removing the clothes hanger. But dividers are best if they're 15 in. to 16 in. wide, which allows enough room to install a closet-pole rosette 12 in. from the wall. In these cases, and wherever else 12-in. shelving rests on top of wider dividers, I cut the dividers at a gentle radius so that they meet the 12-in. shelving for a better look *(Figure B)*.

And I use a small laminate trimmer fitted with a 1/4-in. roundover bit to ease all the sharp edges.

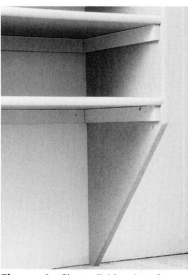

Figure A. *Closet dividers interfere with carpeting and vacuums, so cut them back at an angle near the floor.*

Figure B. *Cut the tops of MDF dividers, too, so that they transit neatly into the narrower upper shelves.*

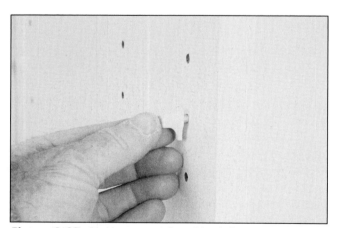

Figure 9-17. *Pegs support adjustable shelves in predrilled melamine dividers. The hole pattern (through-holes and holes for 12-in. and 16-in. shelves) varies from one manufacturer to another, so check the material carefully.*

are manufactured with holes drilled in only one side, other boards are drilled through both sides. Big-box outlets seem to stock dividers that are drilled only for 16-in. shelving; good lumber outlets stock material drilled with an additional line of holes, which accommodates both 16-in. and 12-in. shelving. I prefer to use this material for dividers, so that narrower shelves can be installed above closet rods. Colored pegs fit into the holes and support the shelves *(Figure 9-17)*. However, even when I'm working with adjustable melamine shelving, I often install cleats beneath shelf-and-pole locations as added security, so that the dividers won't buckle and drop the shelves and clothes onto the floor. Unfortunately, that makes the shelf and pole fixed, but if the divider is attached flat up against a return wall or secured to a fixed bank of shelves (more on that in a moment), then cleats aren't necessary and the shelf and pole can remain adjustable.

For shoe shelves and sweater shelves, I make nearly every shelf adjustable — except the center shelf and bottom shelf, which should be secured to keep the dividers from shifting or bowing and releasing the adjustable shelves.

Haefele's Jiffy Connectors® (available at most big-box outlets) are a good alternative to ugly cleats for securing pre-finished shelving *(Figure 9-18)*. These fasteners are handy, quick, and perfect for securing hanging dividers to top cleats — but I'm getting ahead of myself.

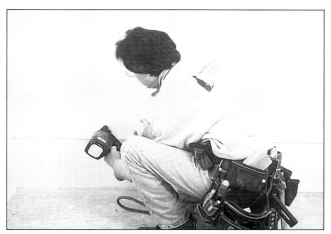

Figure 9-19. *For hanging dividers, install a bottom cleat, or ledger, about 16 in. off the floor. Secure it to the studs, as it will support much of the weight of the shelving and clothes.*

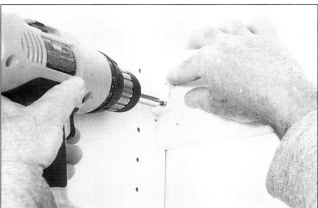

Figure 9-18. *Haefele's Jiffy Connectors® are ideal for fastening pre-finished shelving. First, install the metal under-bracket into the side of the divider, near the center of the top cleat (top). Then install the white cover plate (above). To ensure the screws are secure, always drill pilot holes in MDF and particleboard.*

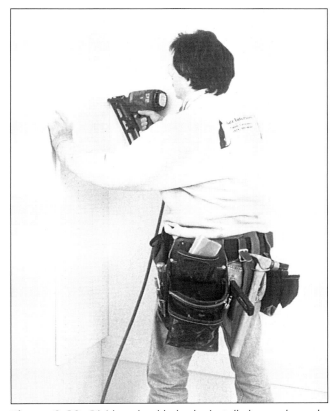

Figure 9-20. *Dividers should also be installed as end-panels so that corner shelves can be adjustable. The end panels give the closet a cleaner look and make the overall installation a little faster.*

Installation

For hanging dividers, always begin at the floor. Mark a level line about 16 in. off the floor for the lower cleat or ledger, and anchor the ledger to the studs *(Figure 9-19)*. Be sure the ledger is well secured to the studs — use 3-in. drywall screws if there's any doubt, because this ledger will support much of the weight of the shelving and the clothes. Lay out each divider on the top edge of this ledger so that all layout lines will be covered by shelving, dividers, or cleats. Rather than drawing plumb lines on the walls for each divider, rely on the top cleats to position each piece (more on that in a moment).

Once the ledger is installed, I fasten uprights on each side wall as end panels, to maximize the amount of adjustable shelving in the closet *(Figure 9-20)*. If the walls are terribly out of plumb, I shim the end panels so that they're parallel with the dividers, and then caulk the gap with white caulk. (To speed production and satisfy my need for perfection, I like all the cleats and dividers to be the same size in each section of the closet.)

Plastic L-clips and drywall anchors will secure the dividers to the wall, and many closet compa-

CLOSET SHELVING 135

Iron-On Edge Tape

All visible cuts in pre-finished shelving and cleats must be edged with iron-on tape. I use a regular household iron to fasten the heat-activated glue *(Figure A)*. I move the iron slowly, so that enough heat is applied to fasten the tape, but not so slowly that the iron burns the material; after all, it's just plastic.

Once the glue has set and the tape is secure, I trim the edges with a double-edge trimmer (see Appendix 1). With two razor-sharp blades, the trimmer *(Figure B)* slices easily through edge banding, whether melamine or wood. After trimming the edge, a little sanding with fine paper finishes the edge perfectly.

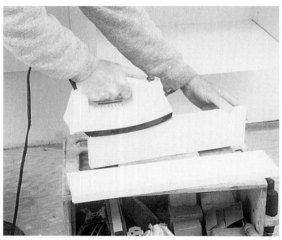

Figure A. *Even though their clothes usually look wrinkled, carpenters use irons too — for iron-on edging.*

Figure B. *This handy banding trimmer saves time and ensures a perfectly flush cut without marring the edge of the shelving.*

Figure 9-21. *Top cleats speed installation and provide more secure fastening.*

nies use them. But I'm not a great fan of drywall anchors — I've seen them fail too often. Instead, I like to install a permanent 1x4 cleat between each divider, flush with the top edge. That cleat is cut the same size as its shelf and helps locate the exact position of the next divider. Also, it never interferes with shelving arrangements and it provides a secure place to mount the dividers and the top shelf — which helps to secure all the shelves to the wall *(Figure 9-21)*.

Next, install any shelves that will be stationary. On adjustable shelving, I prefer to lay out closets with shoe shelves or sweater shelves centered between clothes poles. In that way, central dividers are well secured by fixed bottom shelves, fixed center shelves, or both. Fixed bottom shelves can be nailed to the lower ledger and secured to the dividers, too. I tack the shelf in place, then drill counter-sunk pilot holes for screws *(Figure 9-22)*. Several different types of caps are available to hide the screw head. I've used both snap-on screw caps and stick-on caps with equal success *(Figure 9-23)*.

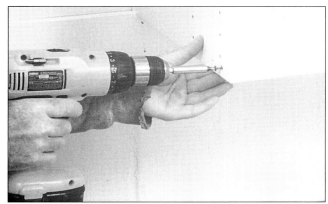

Figure 9-22. *Fasten all bottom shelves with screws, to ensure that the dividers won't spread or move.*

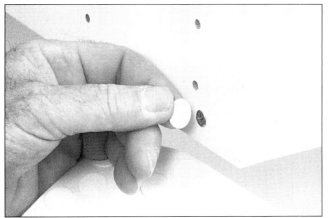

Figure 9-23. *Screw caps, both snap-on and stick-on, are the cleanest way to hide screw heads.*

Figure 9-24. *The top shelf must be fastened to the dividers and to the top cleat, so that the entire unit will be rigid.*

Figure 9-25. *After scribing and cutting one end, measure the wall from corner to corner and transfer that measurement onto the back of the shelf. Then spread the scribes to that mark, and scribe the shelf to fit perfectly the first time.*

Fixed center shelves can be installed at any point. I use screws to secure the top shelf, too, but mostly because I'm always afraid a nail will deflect in the particleboard core and pop out through the melamine skin—just when I'm almost finished with the job *(Figure 9-24)*.

SCRIBING SHELVES

Like hammers left on ladders, challenging jobs eventually fall on every finish carpenter; not every closet is a snap. However, even the most challenging closets—whether they're made with unfinished or prefinished shelving—can be tamed by concentrating on layout and relying on basic carpentry principles.

Simple Scribes

I scribe a lot of shelving to fit irregular walls. If I know the end walls are crooked, I cut the shelf long enough to allow for scribing, then tilt the shelf into position and scribe one end. After cutting the first end, I take a measurement from wall to wall—right at the back corners of the shelf—and then pull a tape from the cut end and mark that measurement on the raw end of the shelf. Finally, I tilt the shelf back into the opening, this time with the raw end down. I spread my scribes to the mark *(Figure 9-25)*, scribe the contour of the wall onto the shelf, and cut exactly to that line. Simple: A perfect fit every time.

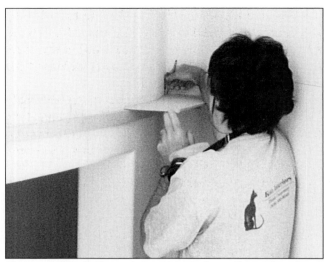

Figure 9-26. *Scribe each template to the radius wall and to the angled corner walls.*

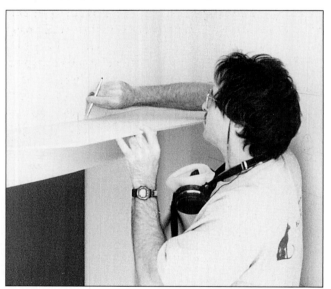

Figure 9-27. *Mark registration lines on the walls and transfer those lines to each template.*

Figure 9-28. *Use the registration lines to ensure that the three templates will combine into one perfectly sized jig.*

Figure 9-29. *Trace the shape of the jig onto the shelf and cut just outside the line with a jigsaw.*

Figure 9-30. *Use a belt sander to fine-tune the cut right to the pencil line.*

Once the side walls are scribed and cut to fit, I concentrate on the rear wall. If the shelf is secured between two dividers, I position the front edge so that it projects an equal amount in front of both dividers, and then spread my scribes to meet the shelf at the farthest gap from the wall. That way, I scribe only enough off the back of the shelf so that it will fit tightly against the wall.

Complicated Scribes

You can't always tilt a shelf into an opening, however, which was the case in a curved closet I did behind a sweeping stairway. Instead, I used 1/4-in. MDF to make templates for the shelving. Because each end was especially vexing (the radius wall died into adjoining walls with acute angles), I made the template from three separate pieces—one for each end and one that spanned the center of the wall *(Figure 9-26)*.

Figure 9-31. *A bearing-guided straight bit and template makes cutting the additional shelves easy work.*

Working alone, it was easy for me to hold the small pieces of MDF against the wall. I carefully scribed each piece until it fit perfectly snug, and left all the pieces long—I had allowed at least 8 in. of overlap.

Next, I made registration marks on the wall and transferred those marks onto each template *(Figure 9-27)*. I used the registration marks to join the three pieces into one continuous jig—precisely the right length *(Figure 9-28)*.

After tracing the shape of the jig onto the first piece of shelving, I cut just outside the line with my jigsaw *(Figure 9-29)*, and then used my belt sander to get right to the mark *(Figure 9-30)*.

Though the jig was carefully constructed, I never imagined that each shelf would fit exactly the same—flat walls are seldom straight and radius walls are never uniform from ceiling to floor. But I did expect that the jig would get me very close. In fact, the variance in the wall from the floor to the top shelf was so slight — with no gap larger than $1/16$ in. — that I decided to turn the rest of the job over to an apprentice. He rough-cut the remainder of the shelves with a jigsaw, and then finished them with a router fitted with a bearing-guided template cutter *(Figure 9-31)*.

CHAPTER 10

Crown Molding

Installing crown molding is one of those magical experiences that finish carpenters are blessed with. This oddly shaped material is considered by many to be a test of any finish carpenter's skill, and for good reason: Crown molding requires all of the basic measuring and cutting skills learned from baseboard installation, as well as a nearly post-graduate familiarity with angles — some of which are imaginary. Though working with crown molding is more difficult than installing baseboard, there are still many similarities, and the corners and joints are described using exactly the same language.

LAYOUT AND MEASUREMENT

If crown isn't installed in a straight line, the molding will wiggle and wobble repulsively from corner to corner, between ceiling and wall, and tight margins between it and tall door and window casings will be hideously unequal. Always snap lines before applying the molding.

To determine where the bottom edge of the crown will sit on the wall, I place a piece of the molding on my chop saw just as it will sit on the wall—with the bottom edge on the saw's base and the top edge leaned against the saw's fence. I rock the crown until both "feet" lie flat against the saw base and fence, and then measure from the fence to the bottom edge of the molding. That measurement is called the rise — the distance from the ceiling to the bottom of the crown.

If I'm trimming only one room, I measure down for the rise and mark the bottom edge of the crown at each corner in the room, and then snap chalk lines across all the walls. (If the walls have already been painted, I use a dry-line and pencil in small marks every 6 ft. or 8 ft. along each wall.) Whenever I'm installing crown in more than one room, especially on large jobs, I transfer the rise measurement to several small blocks of 1x2. Each carpenter working on the job can then carry the same gauge block so that all the crown will be installed evenly (unlike baseboard, installing crown molding is often a two-man job).

To save time and unnecessary steps up a ladder, I measure the length of each piece while I'm snapping lines on the walls. The measurements are always taken right along the chalk line, not at the ceiling. I write those measurements, in large numbers, just above the line on each wall.

To minimize trips to the saw, I make a cut-list for each room. The cut-lists I make for baseboard and crown molding are very similar (see Chapter 8, page 115). The measurement techniques are almost the same, except that I like to cut crown a little tighter than baseboard. When measuring pieces over 10 ft. long, I add $1/16$ in. to $1/8$ in. and spring them into place so that corner joints are tight. For pieces under 10 ft. long, I measure precisely to within $1/32$ in.

The cut-list notations I use for crown molding are also the same as those I use for baseboard: inside corners are noted with a ^; outside corners are marked *OC*; and self-returns are *SR*. For a piece that's 64 in. long, with an inside corner on the left and an outside corner on the right, I write on the cut-list: ^ *64 OC*. If the angle is 90 degrees, then I don't make a notation, but for other angles — like $22^{1}/_{2}$ degrees — I make a note next to the type of cut (see the section of this chapter "Finding the Right Angles," page 145).

CUTTING CROWN

There are two distinctly different methods used for cutting crown molding. One is known in the trade as cutting "in position," the other is called cutting "on-the-flat." Each method has a specific use, and clever carpenters should learn to be proficient in both methods. For each method, I've developed simple routines and techniques that help increase efficiency and accuracy.

Cutting In Position

The most efficient way to cut crown molding is "in position"—with the molding standing upside-down and leaning against the power miter saw fence. Cutting in position means that the material is always in the same position on the fence and never has to be flipped end-for-end; the bevel of the saw is never changed from 90 degrees; and aligning the blade so that it enters the material exactly on the measurement mark is no different—and no more difficult—than cutting baseboard.

Everyone in the trade refers to this method as the mysterious "upside-down-and-backwards" position. No further examination of the enigma seems ever to be required. But carpenters might be better off understanding that the crown molding isn't really upside-down-and-backwards — it's the saw.

Visualizing the correct direction of a cut is easier if we imagine that the fence is the wall but the base of the saw is the ceiling. For inside corners, the long point of the miter and the bevel is always on the wall or fence, just like baseboard *(Figure 10-1)*. For outside corners, the short point of the miter and the bevel is always on the wall or fence, just like baseboard *(Figure 10-2)*. That rule *never* changes.

Whenever I cut crown in position, I always start by placing a short piece of the molding on my saw (upside down). I trace a pencil line across the crown along the fence and along the base. I place a straight piece of scrap material, about 3 ft. or 4 ft. long, at the bottom line and fasten it with screws to my extension wings where it acts like a stop *(Figure 10-3)*. With the stop attached, the molding is held securely in position, the bevel is cut exactly the same every time, and long lengths of material are easy to handle.

Like baseboard, if the piece has an inside corner (most have at least one), I cut that end first, then hook my tape on the long point, and reach across the crown to measure the opposite end *(Figure 10-4)*. With the stop in place, the molding never moves while I'm measuring.

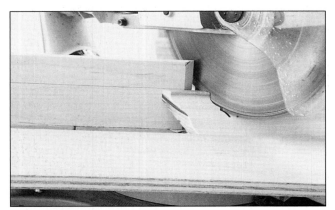

Figure 10-1. *The long point of an inside corner is always against the wall or fence.*

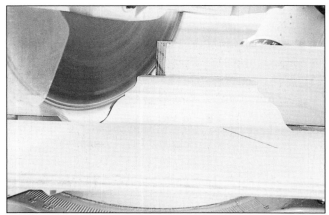

Figure 10-2. *The short point of an outside corner is always against the wall or fence.*

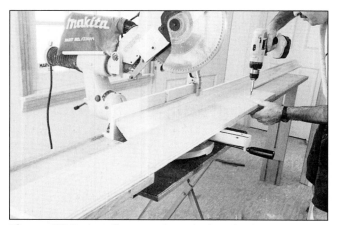

Figure 10-3. *Install a stop along the length of your saw base to stabilize crown molding when cutting in position.*

If the piece has two outside corners, I cut one end and align the short point of the miter with the edge of the fence, then hook my tape *on the fence* and measure to the opposite end of the molding *(Figure 10-5)*. For long lengths, I align the short point with the end of my extension wing, and then hook my tape over the extension wing and measure to the opposite end of the molding.

Also like baseboard, I first cut any end with a self-return, then hook my tape over the long point of the self-return and measure to the opposite end of the molding.

Cutting On-The-Flat

Crown molding with a rise of over 4³/₄ in. can't be cut in position on any 12-in. miter saw. Crown with up to a 6-in. rise can be cut in position on a Hitachi 15-in. saw, which I prefer to use whenever possible, especially on jobs with any real volume, because cutting in position is much faster than cutting on-the-flat. But extremely large crown moldings are becoming more and more popular, and they must be cut on-the-flat with a sliding compound miter saw. Therefore, every carpenter should become proficient at cutting on-the-flat.

Sliding compound miter saws (SCM) have revolutionized more than just crown cutting. Shortly after the introduction of the Hitachi 8¹/₂-in. sliding compound miter saw (the C8FB2, marketed first around 1988), full-size radial arm saws were no longer seen on job sites, and other contraptions — like saw bucks — nearly disappeared. Now, SCM saws are manufactured in 10-in. and 12-in. sizes, too, and unlike the early Hitachi saw, many new models bevel in both directions, a benefit I'll discuss in a moment.

Most carpenters find that cutting on-the-flat isn't nearly as straightforward as cutting in position. I follow a few simple rules that make the job less confusing and more efficient.

Flip the material, not the saw. Though my Makita 12-in. saw (LS1211) bevels in both directions, I try not to change the bevel angle if I can help it. I find it's faster to flip the molding end for end, that is, unless I'm cutting in a small room or hallway. Then I have no choice and have to flip the bevel (more on that in a moment).

Keep the bottom edge measurement in sight. Crown molding measurements are almost always marked on the bottom edge and never on the top. That's one reason why cutting crown in position is easier — the bottom edge of the molding is always

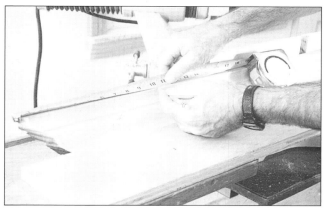

Figure 10-4. *Measure an inside corner by hooking the tape over the long point.*

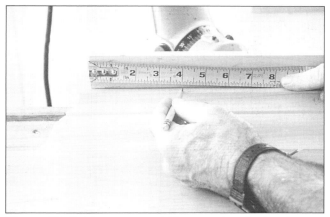

Figure 10-5. *Measure an outside corner by aligning the short point with the end of the saw fence (or the end of the extension wing) and hooking the tape over that end.*

up, so all measurement marks are clearly visible, and aligning the saw blade with the measurement mark is not difficult.

But cutting crown on-the-flat (with the bevel tilted only toward the left) isn't quite so simple because the bottom edge isn't always in view. To cut left-hand (LH) inside corners and right-hand (RH) outside corners, the bottom edge must be against the fence and nearly hidden from view *(Figure 10-6)*. When the bottom edge is against the fence, accurately aligning the saw blade with the measurement mark is not easy; at best it's time-consuming, difficult, and tedious.

If I use a double-bevel SCM saw (and think ahead), I can always manage to have the bottom edge of the molding away from the fence where I can see the mark I have to cut to.

Cut LH inside corners and RH outside corners first. For pieces with inside-to-inside miters, I cut the LH corner first, then flip the molding end for

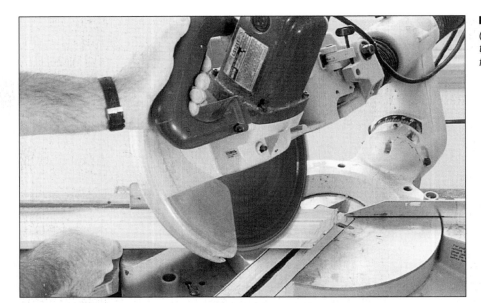

Figure 10-6. *To cut LH inside corners (left) and RH outside corners (below), the bottom edge must be against the fence and nearly hidden from view.*

end and hook my tape on the long point of the miter/bevel to mark the opposite end *(Figure 10-7)*. Cutting the RH inside corner is easy because the bottom of the molding is away from the fence, and I can quickly and accurately align the saw blade with the mark *(Figure 10-8)*.

For pieces that have a LH outside corner and a RH inside corner, I cut the LH outside corner first, and then slide the material along the fence so that I can cut the RH inside corner last. This means measuring from the short point of the miter/bevel, so—like baseboard molding—I align the short point with the edge of my extension wing, and then hook my tape over the wing to pull the measurement for the opposite end.

For pieces with LH inside corners and RH outside corners, for pieces with two outside corners, and for long lengths that are too long to spin in the air *(Figure 10-9)*, I flip the motor to the opposite bevel *(Figure 10-10)*. There's simply no other way to cut these pieces with the bottom edge and the measurement mark away from the fence.

Clearly, a double-bevel saw has several advantages over a single-bevel saw:
- The material rarely has to be flipped or moved.
- Cuts can be made quickly and accurately with the bottom edge of the molding always positioned away from the fence.
- Inside corners can always be cut first, which makes it easier to hook a tape measure.

The only disadvantage is that the saw bevel must be flipped, which requires one more step and a little more time.

Figure 10-7. *On inside-to-inside miters, cut the LH corner first, then measure to the RH corner, hooked over the LH corner long point.*

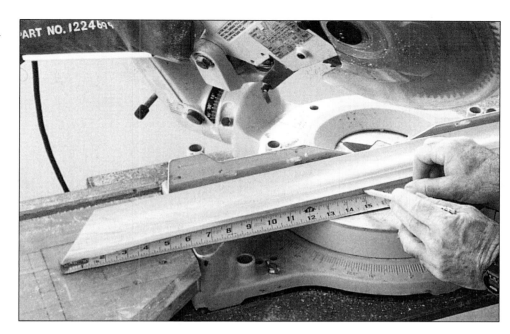

Figure 10-8. *It's easy to see the measurement mark when cutting a RH inside corner on-the-flat.*

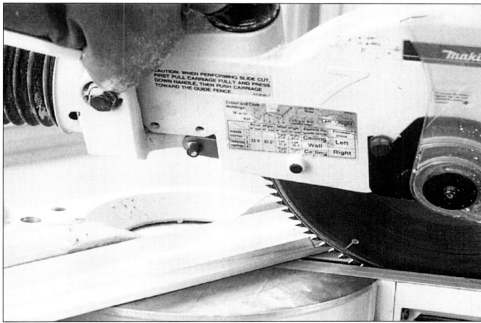

Finding the Right Angles

Cutting crown requires the understanding of four different angles: the corner angle (the angle that the walls meet); the spring angle (between the back face of the crown and the wall); the bevel angle (the angle to be cut, measured from the back face of the crown); and the miter angle (the angle to be cut, measured from the bottom edge of the crown). Bevel and miter-angle charts are available for cutting crown moldings with 45- and 38-degree spring angles (see Appendix 3). But not all patterns of crown molding are milled with a spring angle of 45 degrees or 38 degrees. And the best SCM saw in the world won't be of any help if you can't find the right bevel and miter angles. When I first started using these saws and ran into an oddly angled crown pattern (and couldn't use the standard settings marked on my saw), I always tried to cut the crown in position first, and then determine the right bevel and miter setting from the piece I'd cut.

These days I don't worry about such nonsense because I use a Bosch Angle Finder (BAF) and read the angle of every corner—even the 90-degree corners—before cutting any crown *(Figure 10-11)*. I can't say enough good things about this tool: Simply put, if you can afford an expensive SCM

CROWN MOLDING 145

Finding the Spring Angle

If you don't know the spring angle of the crown you're working with, hold the molding against a wall with the bottom edge down, just as it will mount to the ceiling. Then place one arm of the Bosch Angle Finder (BAF) against the wall, beneath the crown molding, and open the other arm until it's parallel with the face of the molding *(Figure A)*. Subtract 90 degrees from the number on the LED to arrive at the spring angle.

An inexpensive low-tech protractor can also be used to find the spring angle of crown molding. Hold the crown against a wall or flat surface, with the bottom edge down in the same position it will be installed at the ceiling. Then spread the protractor open between the wall and the back of the crown *(Figure B)*. Using this method, there's no need to subtract 90 degrees because the angle on the protractor is the spring angle of the crown.

Figure A. *Use the Bosch Angle Finder to determine the spring angle of crown molding, subtracting 90 degrees from the angle you see here.*

Figure B. *A simple low-tech protractor will also do the job (without having to subtract 90 degrees).*

Figure 10-9. *Long material can be flipped in a large room, but in a small or narrow room, the saw bevel must be flipped instead.*

saw, you can certainly afford a $100 BAF. And rest assured, this tool will instantly shave time from your crown installations and reduce the risk of premature graying and balding.

The BAF is easy to use. Start by spreading the arms on the BAF until the LED reads the exact spring angle of the crown molding you're installing (see "Finding the Spring Angle," then press the black Bevel/Miter (BV/MT) button. The spring angle will be entered into memory — confirmed by the appearance of "SPR" in a black box at the lower left corner.

Next, place the BAF in the corner of the walls and open it until both arms are flat against each wall, then press the same BV/MT button again. The

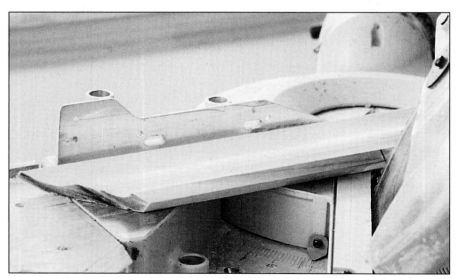

Figure 10-10. *For outside-to-outside corners you have to flip the saw bevel in order to keep the outside edge of the molding (and the measurement mark) visible.*

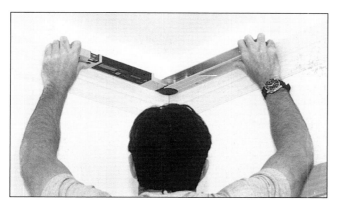

Figure 10-11. *Use a Bosch Angle Finder in every corner.*

abbreviation "CNR" will next appear in the LED, confirming that the corner angle has now been entered.

Press the BV/MT button once more and the miter angle will come up in the display, along with the abbreviation MTR. Finally, press the BV/MT button one last time and the bevel angle will be displayed, along with the abbreviation BVL. It's that simple. When cutting crown on the flat, I read every corner, not just the 22^1/$_2$-degree oddball corners. That way, every joint comes together perfectly tight the first time.

Coping Inside Corners

Coping inside corners has long been the best method of joinery, and for wood crown molding that rule still stands. For MDF and urethane moldings, coped joints are not always the best choice: When MDF is cut to a sharp edge and compressed, the sharp edge folds back like thick skin, and urethane molding manufacturers insist that all joints be compressed and fastened with proprietary adhesive.

Barring certain types of molding, like heavy dentil patterns, and except for those situations where I can preassemble corners (more on that in a moment), I still cope inside corners *(Figure 10-12)*.

Whether using a hand coping saw or a coping foot on a jigsaw, the molding needs to be secure while you work on it. I use my saw table as a worktable, extending the end I'm working on a few inches past the end of the support table and clamping the molding to the fence.

Using a coping saw. Just like baseboard (see Chapter 8, pages 119-120), the first piece into the corner is cut square. I prefer to cope left ends, so I always move to my right when installing crown molding with coped corners (most carpenters quickly develop a favorite end to cope, and it's

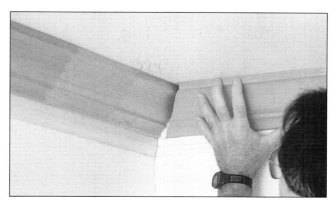

Figure 10-12. *Cope the second piece of crown to match the profile of the first piece (top). Coped joints are tight, lock together firmly, and stay that way (above).*

Figure 10-13. *Begin cutting a cope by making a sliver-thin cut across the bottom of the molding profile. Pull the saw out of the kerf and begin another cut from the back of the molding in order to follow the bottom half of the cove and meet the end of the first cut, releasing the waste.*

usually opposite their dominant hand). I cut the second piece just as I would an inside corner miter, and then follow the profile of the miter with my coping saw. I bring my saw into the bottom of the molding first, and carefully cut the thin point that forms the bottom edge of the corner miter *(Figure 10-13)*. I back the blade out and cut in from the back of the molding to follow along the bottom of the cove until reaching the end of the first cut,

Figure 10-14. *On the third cut (already made in this photo), finish coping the cove profile from the other direction, stopping at the top of the cove, and back the saw out of the kerf (visible here). Remove this waste, again coming in from the back, this time along the fillet that separates the cove from the S-curve.*

Figure 10-16. *Finally, entering again from the back of the molding, cut along the top fillet to remove the waste.*

Figure 10-15. *On the fifth cut, cope the S-curve until the blade reaches the fillet at the top of the molding, and remove the blade.*

which releases the waste. The third cut finishes the cove profile, stopping at the fillet between the cove and the S-curve. I release this waste by cutting in from the back of the molding and along this fillet *(Figure 10-14)*. Then I cut along the S-curve, stopping at the next flat *(Figure 10-15)*. I back the blade out and remove this waste by cutting in along the flat at the top of the molding *(Figure 10-16)*.

Coping with a coping foot. A coping saw works well in small softwood moldings, but a jigsaw, equipped with a Collins Coping Foot® (see Chapter 8, page 120, and Appendix 1) makes quick work out of both softwood and hardwood crown molding. The best person to demonstrate the technique is the inventor of the foot itself, Dave Collins.

Dave starts coping by making a series of relief cuts along each step in the profile. He also cuts a relief at the deepest point in the S-curve. Next, he slices down the S-curve and uses the angle of the coping foot to simultaneously push and leverage the jigsaw blade perfectly along the line of the profile *(Figure 10-17)*. Turning the saw around, Dave then eases the blade up toward the top of the crown and terminates neatly into a relief cut. The bottom of the crown is cut last, first by cutting up through the cove *(Figure 10-18)*, then by cutting parallel with the grain, leaving a final sliver of wood to form the appearance of a miter *(Figure 10-19)*.

INSTALLING CROWN

After all this inside-corner, outside-corner, bottom-edge, cope-or-not-cope business, the subject probably seems a little foggy, but things will clear with a little practice, and applying the molding to the wall is good therapy.

Find the Studs

Before lifting crown into position, I mark the stud layout on the wall, just beneath the snap line and the joist layout in the ceiling, just outside the footprint of the molding. (If the joists aren't running perpendicular to the wall, I cross-nail the molding to the ceiling: Every 18 in. I drive two nails about 2 in. apart and angle the nails toward each other at 45 degrees). I try not to fire a nail unless I know I'm going to hit backing, especially in stain-grade material. Occasionally I use a stud finder (Zircon #50793), though I tend to rely more on a simple magnet, the Tot-Lok Key® *(Figure 10-20)*. The suppliers for both of these tools are listed in Appendix 1.

Don't nail the corners. I usually begin installation with the first piece I measured, especially if I've coped any joints. But I also try to install excep-

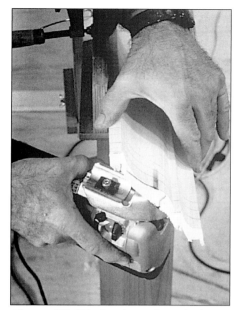

Figure 10-17. *Cut down through the S-curve.*

Figure 10-18. *Cut back up toward the cove relief.*

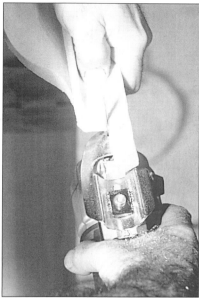

Figure 10-19. *To finish the cut, use a small piece of stock to back-up the fragile bottom edge.*

Figure 10-20. *A magnetic Tot-Lok Key on the ceiling identifies the position of a joist for nailing.*

Figure 10-21. *Use a softwood block to drive the corner together, tapping the top of both pieces toward the wall to close a gap at the top of the miter, or tapping up on the bottom edge of the molding to close a gap at the bottom of the miter.*

tionally small pieces first, just like baseboard, so that longer pieces can be sprung into position.

With a typical two-man crew, molding must be applied to the wall one piece at a time; fastening ends and corner joints must often wait until the next piece is installed. To position the first piece on the wall properly, we carry two short scraps, cut with inside and outside angles. The scraps help ensure that the corners will line up once the final piece is applied. But until that piece is installed, we never nail near the ends or corners—we always fasten the molding only near the center of the wall. Once two adjoining pieces are applied, I work the corner until it's perfectly aligned and tightly closed.

Tap up or down. Working a corner joint together sometimes requires a little patience. Walls and ceilings are never perfectly straight. Miter saws may cut moldings at perfect angles, but a slight bow in a wall, or a belly in a ceiling, will throw a curve into a corner. I carry a block of soft wood so that I never have to hammer directly on the crown. First I tap the bottom edges of both pieces until they're aligned *(Figure 10-21)*. If the top of the corner is open, I tap the top of both pieces towards the wall. As they slide down the wall, the top corner will usually close up. If the bottom of the corner is open, I tap the bottom of the pieces toward the ceiling.

Shim bad walls and ceilings. Wall and ceiling corners are notoriously bad in modern homes. I often have to shim the molding just slightly to

Preassembling Splices

Whenever possible, I prefer to preassemble splices in crown molding, especially for stain-grade material. I make up all splices before beginning the installation, which allows time for the glue to dry a little before cutting each piece for length. Of course, preassembling splices in runs longer than 25 or 30 feet is nearly impossible to manage even for a three-man crew. But for shorter splices I find that preassembling splices is much more productive, and the resulting joints are tighter and far less visible.

I like to use the thickest backing that I can behind the joint because it's easier to secure 3/4-in. backing to the molding than it is to fasten 1/4-in. backing (more on that in a moment). But the backing can't project too far or it will interfere with the crown seating properly against the wall and ceiling. On smaller crown profiles there's often enough room only for thin plywood.

I start by attaching the backing to the back of the inside piece of crown (the piece cut with the inside miter) with glue and staples *(see photograph)*. Next, I glue the joint and close it up tightly at the back of the molding, then fire a few staples through the backing and into the overlapping piece of crown. Before permanently nailing the backing to the crown, I carefully flip the material over and check the joint. With only a few staples fastening the second piece, I can still make

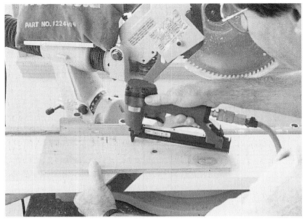

Preassembling splices is faster and guarantees tight field joints. First, fasten the backing to the piece cut with the inside miter, as shown. Then fasten the overlapping piece.

slight adjustments in the fit of the joint. Then I drive more nails through the front (if the backing is thick enough), or more staples through the back (if the backing is too thin for nails).

For large moldings, I prefer to use 3/4-in. backing. With thicker backing, the material can be flipped over before fastening the joint; the joint can be glued and adjusted so that it's completely tight and perfectly aligned; then the fasteners can be driven through the face of the crown into the 3/4-in. backing.

Figure 10-22. MDF shims are often necessary to close up a joint.

Figure 10-23. Caulk between paint-grade crown and wall, as necessary.

close up a joint. For this purpose, I cut shims from scraps of MDF *(Figure 10-22)*. These shims need be only 3 in. or 4 in. long, and 1/4 in. at the heel. Once installed, they snap off easily, just flush or even behind the face of the molding. The shims and the gaps flanking the shims can be covered with caulk.

I try to keep the molding as straight as possible, regardless of the condition of the walls and ceilings, shimming the material wherever necessary to maintain straight lines. If the ceilings or walls are especially bowed on paint-grade moldings, I run a bead of caulk between the molding and the wall *(Figure 10-23)*. I'm a firm believer in filling and fixing unsightly joints and gaps immediately. For

Figure 10-24. *Finish the splice by sanding the seam smooth.*

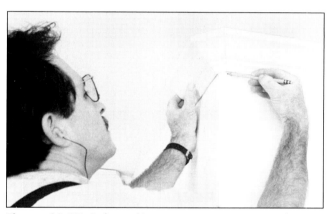
Figure 10-25. *Before taking any measurements, mark bullnose corners with a three-piece mockup.*

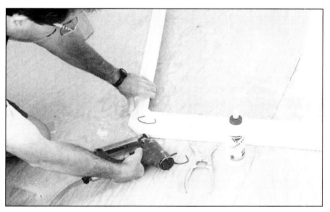
Figure 10-26. *Wherever possible, preassemble complete corners.*

long point against the wall (see "Preassembling Splices"). It's easier to get a tight joint if the second piece overlaps the splice.

I also cut the second piece slightly long. That way the miter can be trimmed slightly at the top or the bottom (the angle of a crown splice can be thrown off if the wall or ceiling has even a slight bow or belly). Even if the splice is cut perfectly, a little sanding is almost always required *(Figure 10-24)*.

Crowning Round Corners

Unlike baseboard molding, premanufactured round corners are not available for crown molding. Instead, crown molding around bullnose drywall corners must be cut at 90 degrees (with the open corner filled by caulk), or a three-piece corner (with $22^1/_2$-degree cuts) must be assembled to complete the corner. I know some carpenters who use a lathe to turn their own profiles for round corners. They drill out the turning so that the back of the profile fits snugly against a bullnose corner, and then they quarter it on a band saw. But that's a method for a different book, not a topic for production finish work.

Cutting a three-piece corner takes a little more time than a caulked 90-degree corner, but I prefer the look because it adds drama to the molding and the corner. Like three-piece baseboard corners, I start by making a corner mock-up, and then use the mock-up to mark every outside corner in a room *(Figure 10-25)*. Most often the small center piece is $3/_4$ in. to $7/_8$ in. wide, from short-point to short-point, with both ends cut at $22^1/_2$-degree angles.

Such a three-piece corner should be preassembled whenever possible (more on preassembly in a moment), so that small pieces don't have to be held and nailed in place. Miter clamps should be used to draw the joints tight *(Figure 10-26)* and hold them that way while they're cross-nailed (see "Using Miter Clamps," next page). Once the corner has been assembled and the glue allowed to dry, installing the three-piece assembly is simple and painless *(Figure 10-27)*.

stain-grade moldings, I leave all gaps open so that irregularities in the walls and ceilings can be floated out by the drywall pickup crew.

Cut splices long. I cut all splices with a compound miter, rather than a simple miter, because a diagonal joint is easier to hide than a vertical joint. The first piece that's installed should always be cut with an inside miter, like an inside corner, with the

Installing Built-Up Crown

Large cornice details are frequently constructed from several layers of molding, along with backing and additional trim moldings. The number of possible designs and combinations of moldings is unlimited, though the examples in the following photographs should provide a starting point for more innovative and clever designs.

> ## Using Miter Clamps
>
> For finish carpenters who care about their work, miter clamps are essential hand tools. They ensure tight fitting miters in casing, crown, and panel molding (more on panel molding and miter clamps in Chapter 11, "Decorative Walls").
>
> I use two types of miter clamps, Ulmia clamps and a lighter-duty set of spring clamps manufactured by the Collins Tool Company (see Appendix 1). The Ulmia clamps come in two sizes and are manufactured from heavy spring steel. These heavy-duty pinchers, with sharp chisel-cut points, grab ferociously and leave a like-size mark. There are times when I find the need for these aggressive clamps, especially when I install large MDF crown moldings. But for most work, the lighter-duty and easier to use Collins Clamps are more suitable. And because they have sharp needle-points, these clamps leave only small marks.
>
>
>
> *Miter clamps: The more aggressive Ulmia clamps are on the left; the Collins spring clamps, which leave smaller marks on the work, are on the right.*

Figure 10-28. *Before snapping lines, be sure the combination of crown and base molding will not interfere with any door or window casings.*

A two-piece cornice molding is often constructed with a layer of baseboard installed beneath the crown molding (refer back to Figures 10-20 and 10-21). The exposure of the baseboard (the amount of the baseboard that is revealed beneath the crown molding) is usually an aesthetic decision, though occasionally other limitations determine that dimension.

Before snapping any layout lines, I check that the two moldings will clear any door or window casings, forced-air registers, electrical boxes, etc. *(Figure 10-28)*. Once the overall dimension of the two stacked moldings is determined, I cut a layout block and mark every corner, then snap chalk lines for the baseboard. After the baseboard is installed, I locate the point where the bottom edge of the crown will rest, and use my scribes to drag a sharp line across the baseboard at that height. A chalk line would work, too, but a scribe line ensures that the crown will be perfectly parallel with the molding details in the baseboard.

Three-piece cornice work is also common. The example in the following photographs uses the same crown molding at the top and the bottom of the build-up. The two pieces of crown sandwich an intervening shelf or center soffit, which also supports the upper member of the cornice.

Always begin a three-piece cornice by building a mock-up of the moldings so that the supporting shelf can be located precisely. Then snap a preliminary line for the shelf or soffit ledger. Before installing the ledger, check along the walls and take several measurements between the chalk line and the ceiling. The chalk line for the ledger and shelf must be located at the closest point to the ceiling; otherwise a bow or drop in the ceiling will

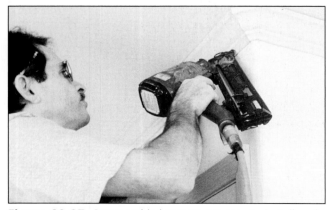

Figure 10-27. *Preassembled corners are easy to install and the joints are sure to be snug.*

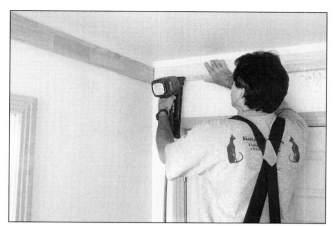

Figure 10-29. *Installing a three-piece cornice begins with nailing the ledger along a chalk line located at the closest point to the ceiling.*

Figure 10-30. *Only the leading edge of this shelf or soffit will be visible, so it can be butt-joined in the corners.*

make it impossible to maintain an even reveal on the shelf *(Figure 10-29)*.

The shelf or soffit is installed next. In this case, only the leading edge of the shelf will be visible between the two crown moldings, so I simply cut the corners with butt joints *(Figure 10-30)*. But the cornice can also be designed so that the bottom of the shelf is exposed for an inch, or even several inches, in which case the corners must be mitered.

Once the shelf is secured, the top crown molding can be installed *(Figure 10-31)*. Because the crown molding must be held at exactly the right reveal on the front edge of the shelf, any large dips in the ceiling must be shimmed, then caulked or floated. The same is true for the lower piece of crown, regarding dips in the wall and the evenness of the reveal on the lower surface of the soffit, especially if the reveal is minimal *(Figure 10-32)*.

Figure 10-31. *Install the upper molding, carefully maintaining a consistent shelf reveal.*

Preassembly

If all walls and ceilings were framed perfectly straight and flat, installing crown molding would be a lot easier—compound mitered joints would pop together just right every time. Imagine that. But the fact is, finish carpenters must perform perfect joinery in an imperfect world. Preassembling crown molding is one way to cope with imperfect walls and ceilings, speed production, and improve the quality of mitered corners *(Figure 10-33)*. Even when installing crown on perfect surfaces, preassembly saves time and ensures better joinery. Besides, measuring, cutting, and preassembling crown molding with absolute perfection is a challenge that I thoroughly enjoy, especially because it saves so much time on multiple-corner pieces *(Figure 10-34)*. I pre-

Figure 10-32. *Install the lower molding, also ensuring a consistent shelf reveal.*

Figure 10-33. *Preassembly guarantees tight-fitting miters, especially with this difficult-to-cope dentil crown.*

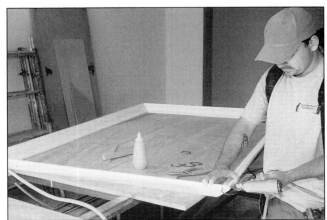

Figure 10-35. *Temporarily secure the corners with miter clamps, and then fasten with glue and brads.*

Figure 10-34. *Measurements must be made carefully, especially for multiple-corner preassembled pieces. But the payoff is increased speed, quality, and satisfaction.*

Figure 10-36. *Small hallway and bathroom rectangles should always be preassembled because they can be installed in just minutes.*

assemble every run of wood and MDF crown molding that I possibly can (urethane moldings should not be preassembled, but more on that in a moment).

Preassembled pieces must be measured perfectly, or labor and material will be wasted on runs that fail the fit test. So I measure all pieces carefully: I measure short pieces with outside corners exactly; for short pieces with inside corners I often subtract 1/32 in.; for long pieces, I add a little, maybe 1/16 in. in 8 ft. The finished assembly must fit tightly but not too tightly. Every carpenter develops his own measuring rules, so if you're just starting out, keep in mind that there's more to carpentry than the numbers you read on your tape measure — those measurements must be interpreted by experience.

Take special care in cutting pieces for preassembly. Be sure your pencil is sharp, your blade is sharp, and your mind is sharp.

I don't cope any inside corners when I pre-assemble crown molding. For a small hallway or bathroom, I miter all the corners at once. I use miter clamps to squeeze the glued-up corners closed, then I secure the miters with brad nails *(Figure 10-35)*. I allow the glue to dry a little before installing the assembly. At first all this special care may seem to take longer, but installation is amazingly quick and painless, and the benefits of careful preparation are soon realized *(Figure 10-36)*.

Installing Urethane Crown

Large built-up cornices, made with multiple levels and several patterns of crown, frieze, and panel molding, can add up in time and material. Urethane crown moldings, manufactured in sizes and patterns that defy the imagination, are fast becoming the alternative of choice. Some of these styles are so large they are impossible to cut even on SCM saws and must be cut in monster miter boxes cobbled

Figure 10-37. *Extra-large urethane crown moldings can be cut using a homemade plywood miter box.*

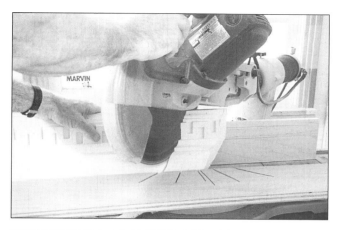

Figure 10-38. *Many synthetic crown moldings must be cut in position (top). If the material is just a little too big for the miter saw, as here, you can use a hand saw to finish the cut (above).*

together on the job site. I witnessed one crew cutting 24-in. exterior cornice molding — in position — against an elaborate 2 ft.-tall and 4-ft.-long plywood fixture using a tree-felling saw. Smaller boxes are easy to fabricate for large moldings that can be cut with a standard hand saw *(Figure 10-37)*.

Synthetic moldings must be cut according to the manufacturer's recommendations: Some types of synthetic molding can be cut flat on a SCM saw, though many urethane styles are manufactured so that they're cut with the face lying flat on the saw (wood crown moldings are cut with the face up). Other types of urethane moldings cannot be cut on the flat at all and must be cut in position *(Figure 10-38)*. Most installation instructions contain other recommendations that are also different from normal applications.

Guard against seasonal changes in size. Urethane moldings will contract during cold months, which is why manufacturers insist that all joints are compressed tightly. The standard recommendation is to add $1/8$ in. for every 5 ft. of length, so that joints will never open. Clearly, these moldings do not lend themselves to preassembly.

In addition, proprietary adhesive (with a urethane base) must be applied continuously along both the top and bottom edges and at all end joints. The ends should be secured first, with the center bowed away from the wall, and then the center sprung into place.

Most manufacturers recommend that splices be cut with butt joints, so that they too can be sprung and compressed together. The importance of compressing the material can't be stressed enough. When installing splices, be certain to add a fractional amount to each piece. I attach each piece at the ends, with a bow in the center *(Figure 10-39)*. Then I use small shims to align the joint. Occasionally the material is manufactured with a few irregularities, and minor filling may be required at both splices and corners *(Figure 10-40)*.

PATCHING CROWN MOLDING

Remodels are notorious for creating joinery problems for finish carpenters. Not long ago, I was asked

CROWN MOLDING

Figure 10-39. *Cut each piece long — even for splices, and bow it into place.*

Figure 10-41. *To patch a section of crown, first cut a short piece of test scrap for each corner. Cut the scrap pieces until the angles match perfectly with the existing molding, and then note the special angle for each corner.*

Figure 10-40. *Minor irregularities in the molding can be filled with spackle.*

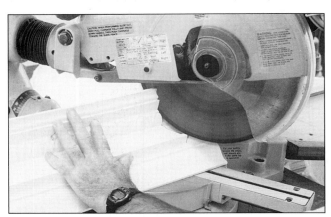

Figure 10-42. *Cutting and trimming custom angles in position is easier, but they can be cut on-the-flat, too. Just note the exact bevel and miter angles for each corner.*

to patch in a section of urethane crown molding around a newly remodeled bedroom wall. The rough carpenters had left some of the old molding on the wall because removing it safely was impossible. The new molding wasn't a perfect match, though it was pretty close.

In order to cut a good tight patch and match the joints, I first cut a short piece for each corner, and then trimmed the scrap piece until it fit tightly against each existing corner *(Figure 10-41)*. I was able to cut this material in position, which made it easier to manipulate the odd angles *(Figure 10-42)*, but I could also have cut the pieces on-the-flat, noting the exact bevel and miter cut for each corner.

After determining the correct angle for each corner, I measured the distance between the short points on the wall *(Figure 10-43)*, and then I cut the piece with the corner angles I had previously noted. I used a pin nailer to help secure the piece,

though a healthy application of urethane adhesive was the permanent fastener *(Figure 10-44)*. Finally, I followed the profile of each corner with a sharp putty knife, removing the excess adhesive and also filling any irregularities between the new molding and the old.

One bit of warning: Urethane adhesive — like roofing mastic — is a mess to work with. I carry a small can of mineral spirits to clean my tools and my fingers.

CUTTING ACUTE ANGLES

As I said in Chapter 7, cutting acute angles used to terrify me. At one time I'd spend hours trying to get the right sized block stuck in behind the crown in just the right spot. But cutting acute angles doesn't have to be a nightmare. With the right jig, these sharply pointed miters can be measured and cut just

Figure 10-43. *Measure the length of the patch piece, in this case from short-point to short-point of both inside corners.*

Figure 10-45. *Acute angles in crown molding are highly dramatic; for me, cutting them used to be dramatic as well.*

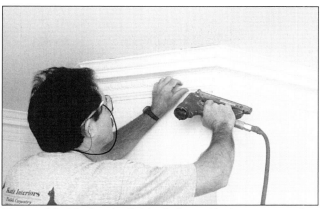

Figure 10-44. *Cut the finished piece with the corner angles previously noted, and it should fit perfectly, especially after the excess adhesive is removed.*

as easily as any standard corner *(Figure 10-45)*.

I prefer to cut crown in position, rather than lying flat, and these angles are no exception. However, it's just as easy to cut the material on-the-flat, and often no accessory fence is required (this depends on the saw, but more on that in a moment).

To cut miters for acute angles with the material in position, I've made two accessory fences that clamp to the extension wings on my CMS. I built the fences 4 1/4 in. tall because that's the maximum depth of cut on my Makita 12-in. saw. Each fence is at a 45-degree angle to the metal fence on my saw. I used a scrap of 3/4-in. plywood, about 12 in. wide and 2 ft. long for the base of the accessory fence. I cut one end at a 45-degree angle and attached a 4 1/4-in. piece of 3/4-in. plywood to the face of the 45-degree cut. I backed up the 4 1/4-in. fence with small triangular pieces, glued and nailed *(Figure 10-46)*. Because the crown molding doesn't have much base support (it projects out the front of the chop saw at a 45-degree angle), I later modified the fence and attached a piece of 1/4-in. plywood to the bottom of each side.

Figuring the right miter angle is no different than cutting casing (see Chapter 7, page 113), but before cutting any material I'm always careful to clamp the molding to the fence. If the molding isn't clamped firmly to the fence, the blade of the saw will pull the material toward the arbor, possibly pulling a hand along with it. Short of that, the cut will stray slightly, and the joint won't be perfect.

Making Sense of a Miter-Saw Scale

To set the saw at the correct miter angle, I simply subtract the miter angle that I want to cut from the angle of the fence, then swivel the saw that sum from the '0' mark. In other words, if I'm working on a 40-degree acute angle and I want 20-degree miters, I subtract 20 degrees from 45 degrees (the angle of the accessory fence). The result, 25 degrees, is the amount that I swivel the saw. (Since miter saw scales mark 90 degrees as '0', it's easy to swivel the saw 25 degrees — that's the 25-degree mark on the saw scale. Remember, miter-saw scales do not represent true angles but rather — from the 90-degree '0' mark — the amount of degrees subtracted from a 90-degree square cut.)

Cutting Acute Angles on the Flat

Cutting on the flat is almost easier, because SCMs can bevel up to 45 degrees, and many can miter up to 60 degrees, well within the angles needed even for a sharp 45-degree corner, a 56-degree miter, and a 46.7-degree bevel for crown molding with a 38-degree spring line.

CROWN MOLDING

Figure 10-46. *An accessory fence for cutting acute angles in position is a must. For safety and tight joinery, be sure to clamp the material to the fence before cutting.*

However, if your saw can't cut a miter that acute, make a simple fence from a wide piece of plywood or 1x12 cutoff. Cut a 45-degree angle on one end and clamp the piece to the base of your saw. Then place the crown molding against the 45-degree angle to increase the miter. Once again, if you leave your miter saw at '0' (90 degrees), the cut will be exactly 45 degrees. If you swivel your saw to 20 degrees, the miter cut will be 25 degrees (45 – 20 = 25)!

Unfortunately, the Bosch Angle Finder isn't programmed to determine bevel and miter angles for corners that are much tighter than 60 degrees, so for most octagons, the BAF is useful only for reading the angle of the corner. To get the miter and bevel angles for a SCM, refer to Appendix 3, "Crown Molding Miter & Bevel Angle Setting," or use the formula provided at www.josephfusco.com/FHB/crownscript.html.

Chapter 11

Decorative Walls

I enjoy installing doors, jambs, casing, base, and crown molding, but like many finish carpenters, I'm always eager to jump into an office, den, library, or living room and get started on the paneling, whether it's simple plant-on trim, wainscoting with recessed paneling and chair rail, or ceiling-height library paneling, with angled bays, intricate detail, and a challenging layout.

In this and the next couple of chapters (covering decorative details for walls, ceilings, and doorways), many of the carpentry techniques discussed in previous chapters will be used for more challenging and rewarding tasks. But regardless of the job's ease or difficulty, remember that careful layout is always the key to a productive and profitable job.

SIMPLE PLANT-ON WAINSCOTING

Wainscoting comes in different models and is known by different names. Because wainscoting corresponds with a classical column and represents the space between the base and plinth of the pedestal (see Chapter 1), wainscoting is also called a *dado*. Occasionally the term dado is used to describe only the flat section of paneling between the top of the base and the bottom of the chair rail.

The simplest type of wainscoting is an inexpensive plant-on look-alike, which, for a modest sum, can have a considerable impact on the decorative appeal of a room. Plant-on wainscoting is simply the installation of small boxes—or panels—of flat molding, fastened directly on the finished wall surface.

Establishing Chair-Rail Height

Unlike recessed paneling, the first step for installing plant-on panels is laying out the chair rail—all other layout lines emanate from the top of the chair rail. On a large job, especially one with many door and window openings, I use a laser level to establish a level line for the top of the chair rail *(Figure 11-1)*. In a small room, I rely on an assortment of levels *(Figure 11-2)*.

Chair rail can be installed at many different heights. The average height is 36 in., but for large rooms with tall ceilings, I've installed the railing at 42 in. and even 48 in. In Victorian and Arts and Crafts-style homes, the chair rail is often installed much higher, with the wainscoting running nearly the same height as the doors. But I've installed the rail as low as 32 in., too, so that the chair rail runs under the stool and acts as a window apron.

Making Sense of the Math

After laying out the top of the chair rail, I measure the length of the wall. To calculate the exact layout

Figure 11-1. *Begin layout with a laser level and mark the top of the chair rail, as all other layout lines emanate from this line.*

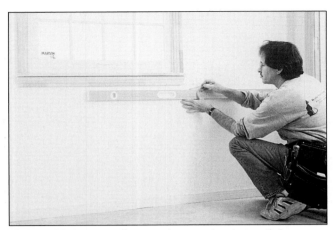

Figure 11-2. *In small rooms, lay out the chair rail the traditional way.*

of the panel spacing, I begin with an estimate of the number of panels. From experience, I've learned that narrow frequent panels increase the feeling of a room's height; alternatively, wide infrequent panels increase the feeling of a room's width. Estimating the number of panels must begin with an idea of the panel width.

On one job a while back, I wanted to increase the height of a low-ceilinged room, so I installed narrow panels and stiles. Also, because I was using slender $1/2$-in.-wide half-round for the paneling detail, I knew the panels had to be even smaller, so I guessed at 10 in. for the panel width and decided on 3 in. for the stile width.

Once the number of panels has been estimated, the exact size of each panel is easy to figure out, but always remember one important rule: There is always one more stile than there are panels — if there are seven panels, then there must be eight stiles (unless the wall you're working on never meets another corner). Therefore, to arrive at the exact panel size, I first subtracted the width of one stile from the total length of the wall (96 in. – 3 in.= 93 in.) See *Figure 11-3*.

For both the math and the layout, it's always simpler to think of the first stile and panel as a single unit, and the same for the remaining stiles and panels. So after deducting 3 in. for the last stile, I next divided the length of the wall by 13 in. The result was $7 1/8$. Of course the number of stiles and panels must be even, so I rounded the number down to the next even increment and divided the length of the room once more—this time by 7. The result was $13 5/16$, which meant that after subtracting the stile width (3 in.), the panel width was exactly $10 5/16$ in.

Layout is also easier if the stile width and the panel width are combined into a single unit of measurement. I laid out the wall by measuring from the corner $13 5/16$ in., and then added that amount to mark the location of each succeeding stile. My last mark was pretty close to 3 in. away from the corner — just enough room for the last stile.

Using a Construction Master Pro Calculator

But "pretty close" isn't good enough for many of the jobs I work on. To speed productivity and ensure accuracy (in other words, to eliminate guessing at the panel size and making repeated layout marks, erasing them and starting over, or circling the "right" ones, and then drawing squares around the "perfectly right" ones) I now use a calculator to determine all layout spacing.

While Construction Master Pro® calculators might seem like a worthwhile investment just for stair-builders and roof-cutters, finish carpenters can also put these mighty yet mini tools to good use. These carpenter-friendly calculators can add, subtract, divide, and multiply feet, inches, and fractions of an inch must faster and more sharply than a carpenter's pencil.

I've found that math problems are best understood with repetition, so some of this example will repeat what I said a moment ago about layout, but the results will be far more precise.

Most walls are rarely an even length and stiles are often a fractional size, too. If the wall in the preceding example measured $94 15/16$ in. long and the width of the stiles was $4 1/2$ in., then the total space remaining for stiles and panels (now I'm using my Construction Master Pro® Calculator) would equal $90 7/16$ in.

To determine the exact size and locations of the panels, divide the remaining length of the wall by the number of panels — if there are six panels, then the result would be $15 1/16$ in. Bear in mind that this figure is a rounded-off number! But more on that in a moment.

The last step highlights the best use of the Construction Master Pro® Calculator. With the location of the second stile (the combined width of the first stile and panel) still visible on the calculator, press the + button and then press the = button. The stile and panel width will then be added together revealing the location of the second stile and panel. For the remainder of the stile/panel locations, just press the = button and each stile/panel location will

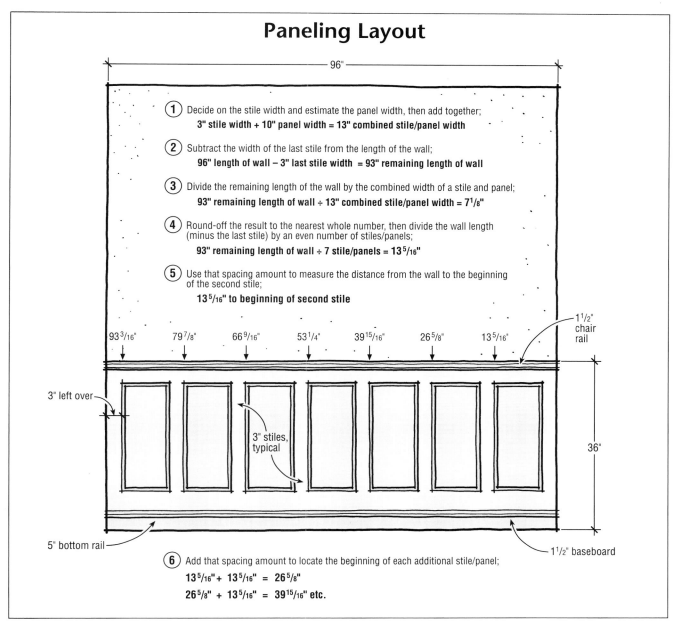

Figure 11-3. *Simplify layout by adding the stile width to the panel width.*

appear on the calculator. However, if you watch closely as the calculator works you'll notice something odd happening. Using this example, the second layout mark would be at $30^{1}/_{8}$ in., the third location at $45^{3}/_{16}$ in., but the fourth location would be at $60^{5}/_{16}$ in. — not at $60^{1}/_{4}$ in. The reason is because $15^{1}/_{16}$ in. was a rounded result and the calculator actually remembers and uses the exact amount. Using a Construction Master Pro to locate stiles and rails in a large room (or for any layout chore from balusters to decking) is the only way to ensure that when you reach the opposite end, the last stile will measure exactly the same as every other stile.

Using a Layout Template

As I've said before, carpentry is often a matter of repetitive tasks. Carpenters who wish to ensure accuracy, reduce the strain of dumb repetition, and improve profit are usually looking for an easier, faster method. For that reason, I use a layout template for most repetitive projects, such as installing plant-on paneling.

Templates are easy to make from scraps of thin plywood. First, cut the template so that it reaches from the top of the chair rail to the bottom of the panel molding *(Figure 11-4)*. Note: If the base hasn't been installed, cut the template a little longer, so that it stretches from the chair rail to the top of the base.

DECORATIVE WALLS 161

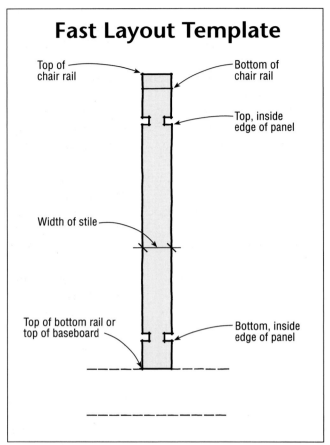

Figure 11-4. *To speed production and guarantee equal-sized panels, use a layout template.*

Next, measure down from the top of the template to the top of the paneling and cut wide notches in both sides of the template — the *bottom* of the notch represents the *top* of the paneling. If the base hasn't been installed and the template is long, measure from the upper notches down to the bottom of the paneling, mark that location, and cut similar notches on both sides of the template so that the *bottom* of the notch is the *top* of the bottom panel.

A slightly longer layout template can also be used to mark the height of the baseboard and ensure that the top of the base is level and parallel with the bottom of the paneling. To use a longer template, simply locate the top of the baseboard on the template, subtract about 1/4 in. to be sure the base will clear any possible hump in the floor, and then cut the template off at that mark.

I hold the template securely against the wall so that the top is flush with the top mark of the chair rail. For laying out the first stile, I also hold a short level up against the template to ensure that my layout lines will be perfectly plumb *(Figure 11-5)*. Then I make a clean, sharp, straight mark at each top notch (for the top piece of molding), a straight mark at each bottom mark (for the bottom piece of molding), and, if the template is long enough to reach the top of the baseboard, I draw a line right across the bottom of the template (in this example the baseboard has already been installed).

I don't need a level to lay out the second stile; instead I cut a spacer block exactly the same length as the width of the panel. I've found that I can bet-

Figure 11-5. *Use a level to make sure the template is perfectly plumb for the first stile.*

Figure 11-6. *Use a spacer block to position the bottom of the template for the second and all succeeding panels.*

Figure 11-7. *Trace the sides of the template onto the wall to mark vertical lines for each panel.*

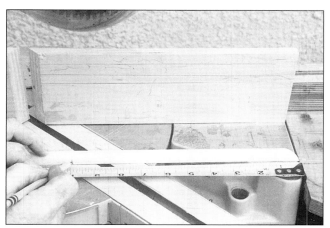

Figure 11-8. *Hook the measuring tape on the edge of the saw to mark perfect lengths of material.*

ter maintain consistent and precise panel sizes by using a spacer block rather than a tape measure *(Figure 11-6)*. And once the first panel is penciled out perfectly plumb, there's no further need for a level. While I'm holding the template, I also draw a sharp line along both vertical edges of the template, for the sides of the panels *(Figure 11-7)*.

The lines on the walls might look a little confusing at first, like shorthand hash marks on a stenographer's pad, but once you start nailing the molding to the wall, you begin to make sense of it.

Cutting Panel Molding

There's always room for improvement when it comes to working with a miter saw — even in this book for professionals — especially if improvement increases productivity. Though some of these instructional tangents might seem like "tips for beginners," there are times when simple answers elude even seasoned craftsmen. So back to basics for just a moment.

Whenever I cut miters, I instinctively ask whether I'm measuring and cutting to the long point of the molding or to the short point. On casing, I'm always cutting to the short point, so I keep the short-point edge of the material away from the fence so that I can enter the saw blade into the material exactly on the mark. For plant-on panel molding, I'm always cutting to the long point, so I keep that edge of the material away from the fence for the same reason.

I start by cutting the miter on the right end, and then lay the material down on the saw table so that the long point is flush with the end of the saw or the end of the table. I hook my tape measure over that end, then measure and make my mark for the left end *(Figure 11-8)*.

After adjusting the material so that the saw blade will enter right on the mark, I set a repetitive stop block, then make the cut. I do the same for the adjacent piece of molding. With two stop blocks, I'm able to cut all the tops and the sides quickly and accurately *(Figure 11-9)*. Also, with two stop blocks, if I make a mistake or lose a piece, I don't have to pull out a tape measure again — both stops remain fixed until all the panel molding is nailed to the wall.

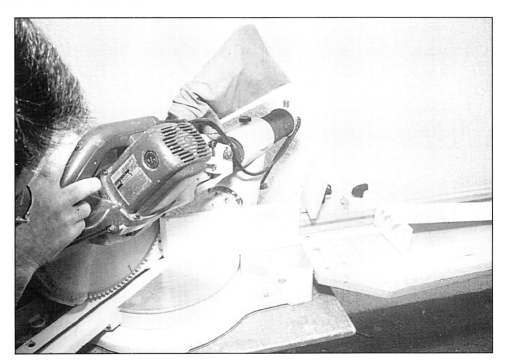

Figure 11-9. *A repetitive stop makes for fast work and ensures that all of the pieces will be exactly the same size.*

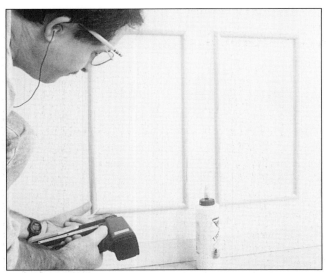

Figure 11-10. *Fasten all the corners but leave the last one loose, and then draw both pieces and the miter together.*

Figure 11-11. *Install the chair rail last, using adhesive, brads, and occasional nails at studs.*

Installing the Molding

I tend to start molding installation at the top, though it doesn't really matter which side of the panels you begin with. After nailing the top in place, I install both legs and match the top miters so that they're perfectly tight. But I fasten only one of the legs all the way to the bottom miter. On the opposite leg I leave the bottom loose. Next I fasten the bottom piece of molding, starting at the leg that's nailed off completely. Only then do I bring the last loose miter together *(Figure 11-10)*.

The chair rail is often the last piece of molding installed. For plant-on paneling, the chair can be attached directly to the drywall with panel adhesive and brad nails, along with an occasional 8d nail into a stud. For stain-grade material, I cope the inside corners; for paint-grade trim, I miter the inside corners as in *Figure 11-11* (more on chair rail later).

RECESSED PANEL WAINSCOTING

Layout for recessed paneling is no different from layout for plant-on paneling: Both types of paneling are governed by the same rules of proportion, and both types depend on panels (or spaces) surrounded by stiles and rails. However, the moldings used to outline recessed panels are often wider than those for plant-on panels, and they also overlap the stiles slightly. For both of these reasons, I increase the stile width for recessed paneling to 4 in., and to $4^{1}/_{2}$ in. or even 5 in. in large rooms.

I should say a few words about outlets, too. I prefer to have electrical, telephone, and stereo outlets placed within the flat panels. However, I've also installed outlets (especially horizontal outlets) in an extended bottom rail. This design is particularly advantageous if the panels are raised, since the clutter of raised panels and outlets can be overbearing.

If the outlets aren't in the right position to miss the bottom rail or an occasional stile, then move them — or insist that others move them. Moving the outlets is not a monumental chore; the paneling will cover all demolition, and the appearance of the finished wall will be correct and proper.

Start at the Bottom

Another difference between plant-on and recessed paneling is the installation procedure. For recessed paneling, I always start at the bottom — no top mark is necessary. These days, I set my laser on a small stack of blocks until the beam is exactly at the height of the bottom line. I'll explain more about the installation procedure in a moment, but remember, the height of the bottom rail must be tall enough to allow for the baseboard molding as well as a minimum reveal of rail before the overlap of the panel molding.

Laying Out Stiles and Rails

The rails can be cut first, and the stile layout made directly onto the rails, but I also mark the stile layout on the walls. Having the walls marked, at least along the bottom rail (especially on long walls with spliced rails) is the best insurance against math mistakes. Corners pose special problems —

Inside and Outside Corners

Inside corners are usually blind. In other words, the ends of the rails are hidden in the corner, and so are the ends of the rails on the adjoining wall. But outside corners are another story and require a little extra attention to layout.

To understand paneling, it's best to imagine the stiles and rails on a door: the stiles run through to the top and bottom of the door so that the ends of the rails won't be visible. I design all outside corners the same way — so that the outside corner pieces are continuous from the floor to the chair rail. In some cases, I miter the outside corners, too, especially for paint-grade and stain-grade material made with an MDF core. However, I butt the corners and highlight the joinery for outside corners in Craftsman-style homes.

Mitering outside corners looks difficult, but it isn't *(Figure A)*. I join the mitered pieces before installation and roll the long-point edges slightly with a block of wood to ease the edge and close the miter tightly. After the corner is done, I install the bottom rail, and then the top rail. I install the field stiles last by sliding them over the splines between the top and bottom rails.

Occasionally, outside corners occur on narrow return walls and must be scribed to an adjoining wall *(Figure B)*. The first step is to cut the panel slightly oversized, allowing enough additional material for the scribe. Next, temporarily fasten the panel to the wall (one or two brads work well for this task and leave almost invisible holes), but be careful to keep the front edge and miter parallel with the outside corner. Measure the distance that the short point of the miter projects past the outside corner, then spread your scribes that exact amount, plus a little bit extra for a fudge factor. Then scribe the back of the piece to fit the shape of the rear wall.

Figure A. *Install outside corners first, then the bottom rail, top rail, and backing for the base. Install stiles in the field last.*

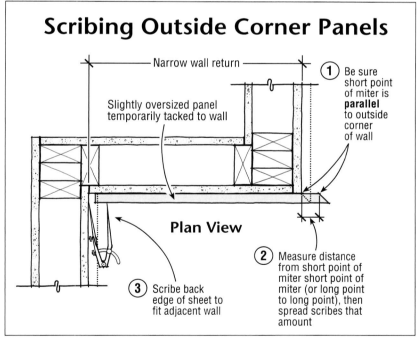

Figure B. *When scribing a panel for a narrow return wall, cut the panel oversized, temporarily tack it in place, and set the scribe to match the distance the miter's short point overhangs the corner.*

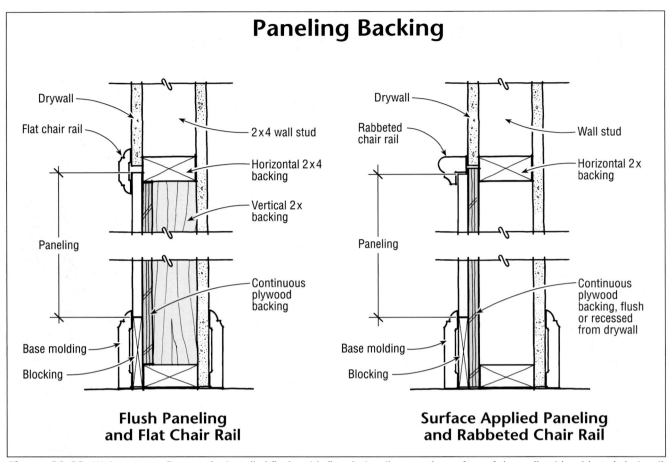

Figure 11-12. *Wainscot paneling can be installed flush, with flat chair rail, or on the surface of the wall, with rabbeted chair rail.*

particularly outside corners. I always lay out the corners before I jump into the deeper water of cutting and installation (see "Inside and Outside Corners"). Once I've marked the layout on the walls, I cut the top and bottom rails, pair them together, and mark the layout again, across both rails.

Joining Stiles and Rails

Building recessed paneling right on the wall is a common practice in my area, especially in large rooms with long walls — which would otherwise require awkward assemblies and frequent splices — but there are times when preassembling the panels is easier, especially when there is space to set up a comfortable job-site shop. (See "Preassembling Stiles and Rails.")

Over the years, I've seen and used an assortment of techniques for field-joining stiles and rails. Some have worked well and some haven't. Originally, I followed the method of an older, more experienced carpenter: I used small shims to flush up the glued joints between stiles and rails, and I used panel adhesive to permanently secure each piece to the drywall. Unfortunately, I discovered that changes in humidity caused the components to swell and shrink, and eventually caused the joints to fail.

Continuous solid backing does help secure paneling, but I rarely work in a home where the decorating schemes and paneling locations are finalized before the walls have been primed. If the paneling locations are located first, these areas can be sheeted with plywood matched to the thickness of the drywall ($5/8$ or $1/2$ in.), or in cases where the paneling must be flush with the drywall so that the chair rail isn't proud of the casing, blocking can be installed in the stud bays and plywood nailed flush with the face of the rough framing *(Figure 11-12)*.

Biscuits aren't always better. Even without continuous backing, I've found a technique that guarantees tight-fitting, perfectly flush joints that never move—and I don't use biscuits. Though they create a durable long-lasting joint, I've never been able to rely on biscuits to register two pieces of material perfectly flush, mostly because the kerf is oversized slightly to allow room for the biscuit to swell. Even a small amount of joint movement is unacceptable with MDF-core material because the hardwood veneer is too thin to allow much sand-

Preassembling Stiles and Rails

Stiles and rails can also be preassembled into large panels. I find this alternative profitable only on jobs with short runs that don't require splices. However, installing preassembled panels can certainly save time on the job site and allow more work to be performed in an often more productive shop environment.

The spline method I use can also be adopted for shop-assembled panels, but pocket holes are much faster, ensure tight-fitting, perfectly flush joints, eliminate the need for any clamps, and allow non-stop production panel assembly.

The Kreg Jig® has changed many aspects of cabinetry and carpentry, and made face-frame assembly (among other things) more accessible to all carpenters. I use the K2000 ProPack® (see Appendix 1) on a host of different types of work — from assembling 1 1/2-in.-thick bookshelves (see Chapter 14, "Bookshelves") to building 1/2-in.-thick drawers, and in this case, preassembling stile-and-rail paneling.

I drill all the stiles first, both top and bottom *(Figure A)*. I glue-up the frame on an assembly board made from a 1/4-in. plywood back screwed to 3/4-in. by 4-in. rails *(Figure B)*. The width of the rails allows me to access the joint with Kreg's face clamps. A 6-in. face clamp comes with the ProPack, but 10-in., 11-in., and even larger clamps are also available. Once the clamps are installed and the pieces secured flush, all that remains is to drive in the screws *(Figure C)*.

Preassembled frame-and-panels are easy to install. I set the frame on the floor and shim the bottom until it's level and horizontally aligned with the previous section at the corner *(Figure D)*. Because the two frames overlap in the corner, almost any amount (up to 3/4 in.) can be scribed off the second frame and the resulting corner will still look good. But a tight fit is most important. I spread my scribes about 3/8 in. and make the cut with my small Makita panel saw.

Figure B. *Assemble the frames on a glue-up board, using Kreg's deep-throat 6-in. clamps to keep the pieces flush.*

Figure C. *The last step is to drive the pocket-hole crews.*

Figure A. *The Kreg Jig® makes quick, accurate work of drilling pocket holes.*

Figure D. *Shim the preassembled panel so that it's flush with the top of the previous piece, and then scribe it to fit the corner.*

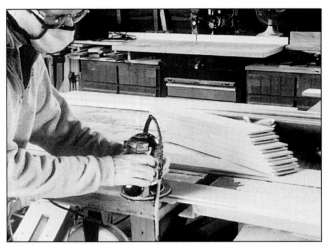

Figure 11-13. *Use a bearing-guided slot-cutter to cut 1/4-in. kerfs at each stile location on the top and bottom rails.*

Figure 11-14. *The first stile must be perfectly plumb.*

Figure 11-15. *All corner stiles must fit tightly against the wall. I spread my scribes the width of the gap, and then trace the contour of the wall onto the stile and cut along the line with my panel saw.*

Figure 11-16. *Splines guarantee tight-fitting, perfectly flush, long-lasting joints.*

ing. In addition, I've found it's easier to hold a router flat and steady on the material than it is to use a biscuit machine — which tends to shift and wobble just at the wrong moment and throw off joint registration.

Instead of using biscuits, I cut spline joints with a router and 1/4-in. slot cutter *(Figure 11-13)*. The bearing-guided cutter I use makes a 1/2-in.-deep kerf, so for splines I cut 7/8-in.-wide strips of 1/4-in. MDF. These 1/4-in. splines fit snugly in the kerf, so that the joints are *always* perfectly flush. For standard 4-in. stiles, I crosscut the splines to 3 1/2 in. wide, so that they don't project beyond the stiles.

Installing Stiles and Rails

I assemble and install most wainscot stile-and-rail paneling piece-by-piece, right in the field, and I always start with the bottom rail. I spread panel adhesive on the back of the bottom rail, and then nail it to the wall, locating the studs with a magnet as I go.

The first stile must be plumb. Because I use a production, stop-block system for installing the remaining stiles and for repetitively cutting all panel molding and panels, I check that the first stile is perfectly plumb *(Figure 11-14)*. If the first stile isn't plumb, the appearance of the first panel will be affected, as will the spacing to the next stile, which hinders repetitive cutting.

In many cases, the first stile must be scribed to the wall, and in rare cases (especially when the first stile is not partially covered by additional paneling) the first stile must be cut wider to allow for scribing. But scribing is easy and doesn't require any math. I hold the stile plumb against the wall

Figure 11-17. *Wipe off squeeze-out quickly with a wet cloth, especially on stain-grade material.*

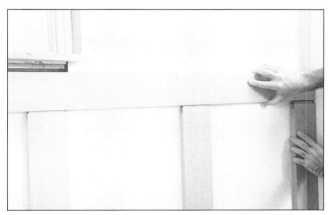

Figure 11-18. *Align the top rail over each spline, then compress the joints with nails or shims before fastening the top rail to the wall.*

and spread my scribes to match the widest gap between the plumb stile and the out-of-plumb or uneven wall *(Figure 11-15)*. Then I trace the contour of the wall onto the edge of the stile and cut along that mark with my panel saw (see Chapter 3), back-cutting the edge as I go so that nothing will hinder a tight fit to the wall.

After scribing the stile, I spread glue in both kerfs and insert the spline in the bottom rail *(Figure 11-16)*. I also spread a little panel adhesive on the back of the stile, then place it on the wall and press it tightly against the bottom rail *(Figure 11-17)*. Once the joint is tight, I nail the stile to the wall, but I nail only near the edges, where the panel molding will later cover the nail holes. I fasten the corner stiles completely to the wall, and then — while the glue is still wet — I wipe off any excess with a wet rag, especially on stain-grade material.

Once the corner stile is installed, the process speeds up considerably. For stiles in the field, I glue the spline joint, spread a little adhesive on the back of the stile, and then press the stile and rail joint tightly together. I nail the bottom snug to the wall, and — because I don't always mark the layout on the wall but on the top rail — I use a spacer-block to locate the top of the stile.

The spacer block is exactly the width of the panel and guarantees that all the molding and the panels can be cut without having to pull out a tape measure, reset a table-saw fence, or change the repetitive stops on a chop saw. However, even when using the spacer block, tack the top of the stile to the wall with only one or two nails, so that it's easier to engage all the spline joints while installing the top rail. Install the top rail as soon as possible, definitely before the glue and panel adhesive begins to dry, but only after all the stiles are on the wall, or at least after all the stiles are within a spliced section of railing.

Glue the kerf on the top of each stile, insert the spline, add a little more glue on top, and then slide the top rail over the splines. With a little jiggling and occasional help from a small prybar, the top rail will slide over the splines *(Figure 11-18)*. Press the rail down tightly against the stiles, but only tack it in place. Check that the stiles are positioned near the layout marks on the top rail before fastening everything permanently.

If the joints aren't tight enough, I drive my prybar a little into the drywall above the top rail and pry the rail down to close up any errant seams. To further squeeze and secure the joints while the glue dries, small shims can also be driven right into the drywall — the chair rail will cover any scars.

Cutting the Panel

Rectangular panels are easy to cut, especially when nearly all of them are exactly the same size. I use my table saw to rip the material to the panel width first, and then I set up the fence to make the crosscuts (see "Tools for Cutouts," next page). I cut the panels exactly 1/8 in. short of the inside dimension, so that they'll all fit without a struggle.

Notches for electrical, stereo, and telephone outlets are common in wainscoting panels. I make a habit of drawing a little rectangle or circle on my notepad for each panel that requires a cutout. I make the drawing with the lines on my pad representing the grain in the wood (more on that in the next chapter). I measure from one stile and one rail

> ## Tools for Cutouts
>
> I use several different tools for making cutouts in panels, cabinets, and countertops. A jigsaw has been and always will be indispensable. I also use my small Makita panel saw for making longer straight cuts. Sometimes I drop the panel saw blade straight into the work for each side cut and finish the job with the jigsaw.
>
> But new tools are constantly changing my work habits, and now I often use a RotoZip, particularly for cutouts in 1/4-in. panels. The RotoZip, equipped with a sharp spiral cutter, cuts smoothly if slowly. The noise is a little more disconcerting than my panel saw, but the little light on the head of the tool overcomes the strain on my eyes, and the small blower gently removes all sawdust — away from my face. And the circle-cutting attachment works like a charm *(see photo)*.
>
>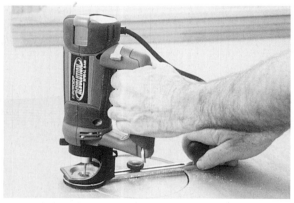
>
> *The RotoZip cuts smoothly, if slowly, and is often my tool of choice for cut-outs in 1/4-in. panels.*

Figure 11-19. *Mark a large X on the back of the panel to ensure that layout for outlets, lights, and speakers will be correct.*

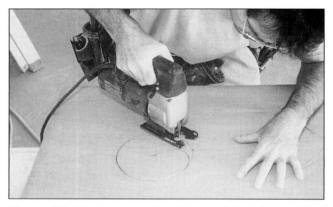

Figure 11-20. *Lay out and cut the panel from the back to eliminate tearout.*

to the offending outlet and subtract 1/16 in. because I always undersize my panels by 1/8 in. These rules are easy to remember and the math is even easier to calculate.

Whether the cutout is for an outlet, switch, speaker, or light, always lay out for the cut on the back of the panel, never the front *(Figure 11-19)*.

Draw an X on the back of the panel if it helps you to remember which is what. That practice will ensure that cutouts are never made in reverse, and cutting from the back of the panel will also eliminate tearout on the face of the panel *(Figure 11-20)*.

Cutting the Panel Molding

Here's one aspect where recessed paneling truly profits from production techniques. With the right jig, this repetitive chore can be assigned to even a green apprentice, as long as he or she has learned how to use a chop saw safely.

Cutting rabbeted panel molding is a little confusing, mostly because the true short-point (at the long-point shoulder of the rabbet, not the long-point of the molding) isn't visible where the blade enters from the top of the material. But measuring, marking, and cutting these pieces needn't be a struggle. A cutting jig removes all the strain and simplifies the task immeasurably.

Make a jig. Most of the panel molding I install is cut with a 1/2-in.-deep rabbet, so I use a scrap of 1/2-in. plywood for the cutting jig (if the molding has a 3/4-in. rabbet, then use 3/4-in. material for the jig). I place the piece of plywood (a 4-in.-wide by 3-ft.-long scrap works fine) up near the fence on my chop saw and leave just enough room for the molding to slide between the jig and the fence — that's with the rabbeted edge of the molding facing me and lipped over the jig. The jig positions the molding so that every cut is made at exactly the

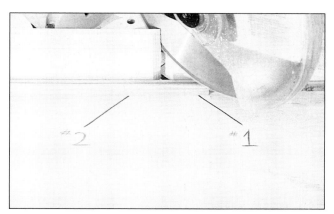

Figure 11-21. *Use a cutting jig to accommodate the molding's rabbet, and label the cuts #1 and #2.*

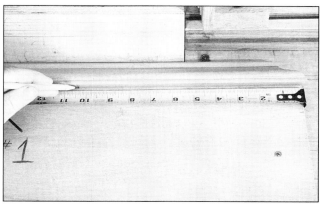

Figure 11-22. *Flush the short point of the rabbet with the edge of the cutting jig, and then hook a tape measure on the edge of the jig to measure the first piece.*

Figure 11-23. *Align the pencil mark with the edge of the kerf and set a repetitive stop for that length of molding.*

same angle and all the miters fit together precisely.

Once the jig is screwed down to the wings on my saw, I cut the miter on the right end of the molding. The kerf that's cut in the jig is marked #1 as in *Figure 11-21* (labeling the jig eliminates confusion for beginners and allows busy professionals to think about other things). Then, I swing the saw to the opposite miter and make a second cut — but this time just in the jig. This cut is marked #2. As you can see, the jig isn't too elaborate.

Simplify the math. The rabbet on most panel moldings is between $5/16$ in. and $3/8$ in. wide. When marking a short-point measurement on the face of the material, the rabbet width must be added to the length of the material — twice for each end if you're measuring from the short point on the face of the material. Figuring the math for each cut is too time-consuming and confusing for me, so I also use the cutting jig to help measure the first piece; I use repetitive stops for succeeding pieces.

Use your jig. I place the short point of the rabbet (remember, it's beneath the molding and is not the short point of the molding) flush with the edge of the jig, then hook and pull my measuring tape from the same edge *(Figure 11-22)*. That way, I have to add the width of the rabbet only once. I mark that measurement, and then carefully align that mark with the #2 in the jig — the kerf ensures precise accuracy *(Figure 11-23)*.

Check and repeat. After making the cut, I check that the piece is exactly the size I need. Then I set it back on the jig with the blade up against the cut, and position a flip-stop for that length of molding. Finally, I cut a new piece, held snugly against the flip-stop, to be sure the setup is perfect — before cutting 30 or 40 other pieces (or sometimes a lot more!).

Use your material efficiently. I repeat the same procedure for the second length of molding, and set a second flip-stop so that both pieces can be cut repetitively and simultaneously. It's often most efficient to vary the sizes that are cut from a single length of material: Sometimes I get three pieces of the first length (and one or two of the second length) from each stick of material, with a minimal amount of waste. *Note:* Only two sizes of molding are needed for most panels, but I'll cover multiple sizes and dog-leg panels in the next chapter.

Assembling the Panel Molding

There was a time—and it wasn't so long ago—when I was unschooled in the fine art of assembling panel molding. I didn't know about miter clamps and pre-assembly, and before I met Gary Ashburn and Jed Dixon, I always struggled to assemble the molding

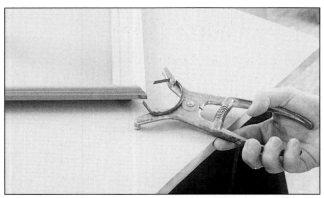

Figure 11-24. *Load the miter clamp in the wrench before gluing up the miter.*

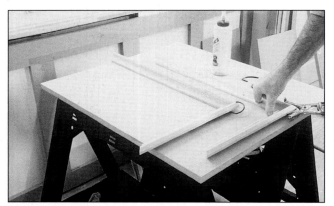

Figure 11-25. *Turn the wrench upside down and apply the clamp under the molding onto the shoulder of the rabbet.*

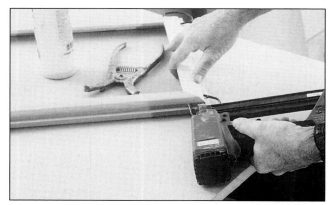

Figure 11-26. *Once all the corners are clamped, shoot brads through the miters right at the shoulder of the rabbet where the material is often the thickest.*

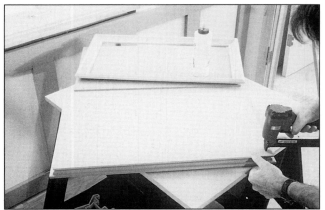

Figure 11-27. *Attach the panel to the back of the molding so that irregularities in the wall won't interfere with the molding lying flat against the panel.*

on the wall, one piece at a time. Success and pleasurable joinery were rare. The first miter always went together smoothly, and sometimes even the second miter looked all right. But when I reached the third miter — and definitely by the time I closed the rectangle at the fourth miter — the joinery was terrible. I've since learned how to avoid that scenario.

Now I attempt to preassemble every stick of panel molding, even for intricate dog-leg panels and acute-angle trapezoids. A flat work table, a bottle of glue, a brad-nailer, and a set of miter clamps are the only tools necessary to create pleasure from what would otherwise be pain.

I start with a clamp loaded in my wrench, and then I glue up each side and press the miter together; holding the two pieces firmly flat against the work table helps to register the miter *(Figure 11-24)*. Then I squeeze the wrench, turn it upside down, reach under the miter to the rabbet, and release the clamp and all its stored pressure. Each time the same thing happens: The miter closes up perfectly tight and a thin line of glue rises from the joint. A little experience taught me to spread just the right amount of glue so that wipe-up is quick and easy. To avoid gluing the one leg to another leg or gluing two heads together, I begin by joining two identical sets — one leg, one head *(Figure 11-25)*. Then I join these two sections in the finished panel.

Once all four miters are assembled and clamped (see "Collins Miter Clamps"), I turn the molding over and shoot a brad through each corner, right near the outside of the rabbet where there's lots of meat *(Figure 11-26)*. I install the flat panel next, to further reinforce the miters.

Attaching the Molding to the Panel

The one disadvantage to using a cutting jig and preassembling all panel molding is that all the joints are cut exactly the same and will not tolerate any bow or belly in the wall. Irregularities in the wall will draw the molding away from the flat panel.

To avoid this problem, and to speed up produc-

Figure 11-28. *Apply a few balls of adhesive on the back of the paneling, and then tilt the panel into the opening.*

tivity, glue and staple the flat panel to the back of the panel molding first. Spread a little glue on the back of the molding, and then place the panel on top. Because both the panel and molding are cut exactly the same size, it's easy to line up the edge of the panel with the rabbeted shoulder of the molding *(Figure 11-27)*. Use a short staple to secure the panel to the molding, and then remove all the miter clamps and move on to the next panel.

Installing the Panel

After the panels are completed, start with an air hose stretched to the far corner of the room. Spread a few marble-sized balls of adhesive on the back of the panel, and then tilt the panel into the opening *(Figure 11-28)*. Because the panels are slightly smaller than the opening, be sure to align the panels across the top, though large irregularities are unlikely—this system is dependent upon a straight line at the bottom rail and perfectly sized stiles cut with a repetitive stop.

Fasten the panel molding to the stiles and rails, and then press the panel against the wall gently and evenly so that the adhesive will spread and eliminate any bumps. It isn't usually necessary to shoot even one nail through the panel itself.

RECESSED PANELING ON RADIUS WALLS

Paneling on radius walls always looks provocative and, like radius doors and casework (see Chapter 13, "Decorative Doorways"), it highlights the talents of a finish carpentry crew. But most of us who have worked on radius walls know that it's not as difficult as it seems. Once again, careful layout and simple techniques make all the difference between a dream job and a nightmare.

Whether the material is paint-grade or stain-grade, veneered MDF-core material can still be used for the panels and the stiles and rails. MDF-core material, especially in 1/4-in. sheets, bends to almost any radius without difficulty. If the radius is especially tight, the top and bottom rails can be relieved with multiple kerfs, which will turn any piece of wood into spaghetti. The relief kerfs are later covered by panel molding.

> ## Collins Miter Clamps
>
> I use two types of miter clamps for preassembling moldings. I usually depend on Ulmia clamps and their awesome power for larger moldings, like heavy crown and base. For lighter pieces, I prefer spring clamps, like those manufactured by the Collins Tool Company (Appendix 1). These lightweight clamps *(see photo)* have plenty of power for most moldings — even most crown moldings — and they also have
>
>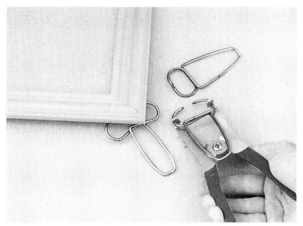
>
> *Collins Clamps® have needle-sharp points, which leave innocuous marks unnecessary to fill.*
>
> a wider range of adjustment, so they're available in only one size. The points on the Collins Clamps® are needle-sharp, unlike the chisel-points on the Ulmia clamps. I prefer the sharper points because they leave a smaller mark in the material, more like a pin-nail hole, while the chisel-points leave a dent that requires proper filling.

Paint-Grade Curves

For paint-grade paneling, I use flex-trim panel molding (see Chapter 2). I cut the pieces long and tack them in place with just a few nails so that I'm sure they're exactly the right length before fastening them permanently to the wall. (No, I've never been able to figure out a foolproof way of pre-assembling panels for radius walls!)

Stain-Grade Curves

For stain-grade material, the challenge gets a little bit tougher. Few budgets can afford the expense of custom radius millwork. So, like many carpenters, I try to make do with the material I have. I order extra material for radius walls, knowing I'll be wasting some of it with experimentation.

Stain-grade panel molding cannot be kerfed or relieved so that it will bend around curved walls, but it can be ripped *(Figure 11-29)*. I start by resawing the panel molding on my table saw. First I cut the top of the material free, right at the shoulder of the rabbet. The top is usually only 3/8-in. to 1/2 in.-thick and will then bend easily around most curves. However, the bottom can be more troublesome, especially on larger moldings with a 3/4-in. rabbet. So if necessary, I rip the molding again, picking a spot where a raised or round detail meets a flat section in the profile.

I reassemble the molding right on the wall, using nails, glue, and sometimes screws to pull each layer tight against the previous layer *(Figure 11-30)*. Wherever my saw blade cuts, I also add a thin layer of solid stock, so that the profile of the horizontal molding will match the solid vertical pieces. After installing all the pieces, I sand the profiles by hand or with a Porter Cable profile sander (see Appendix 1).

FULL-HEIGHT LIBRARY PANELING

The techniques I've described for installing simple recessed paneling are exactly the same as those used for more complex library paneling, even if the panels extend entirely to the ceiling *(Figure 11-31)*. Layout is the only difference, and for all successful jobs, careful layout is critical. (Layout techniques for paneling will also be covered in Chapter 12, "Decorative Ceilings.")

For full-height library paneling, I plead with the owner, decorator, or GC until I have a piece of the baseboard, the casing, and the crown in my hand. Then I begin a careful layout: There are a lot of layout details to consider and overlooking even one can spell disaster.

Casing Details

For full-height paneling, I install or build out doorjambs to the thickness of the finished wall—

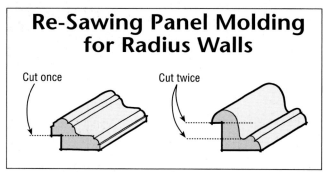

Figure 11-29. *Rabbeted panel molding is easier to bend if it's resawn first.*

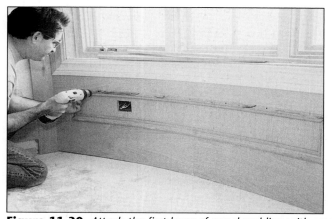

Figure 11-30. *Attach the first layer of panel molding with screws, then draw it tightly against the paneling and the rails (left). Tack the second (top) layer at the end and hold or tack the piece in place to check the opposite miter for length. With a scrap of panel molding ensuring that everything fits right, permanently nail the top layer onto the bottom layer (right).*

Figure 11-31. *The techniques for complex paneling, such as this full-height library installation, are basically the same as those for simple paneling.*

including the stiles — so that the casing can be applied on top of the stiles. I increase the width of all stiles that abut doorjambs to allow for the width of the casing and a reasonable reveal — 2 in. is minimum *(Figure 11-32)*.

Electrical Details

I handle electrical outlets the same as I do for simple wainscoting and try to place outlets in the panels or extend the bottom rail to encompass them. Light switches, however, are another story, especially long gang-boxes with multiple switches. These monsters are too long to fit into a vertical stile, so I try to have the GC inform the electrician to hold them back from the rough opening about 12 in., so that they'll land in a panel. Unfortunately, that's not always possible, and occasionally I'm forced to box around switches with stile material and panel molding. The result isn't really so bad *(Figure 11-33)*.

Narrow Return-Wall Details

Walls alongside doorways, mantels, and windows are often too narrow for recessed panels, so I install a solid panel in those locations. However, I prefer to squeeze in a recessed panel if at all possible. In my opinion, if the exposed recessed panel can be at least twice the width of the panel molding, it passes

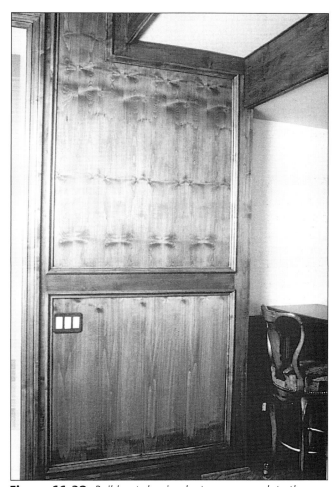

Figure 11-32. *Build out doorjambs to accommodate the thickness of floor-to-ceiling paneling.*

Figure 11-33. *Electrical outlets and switches are best positioned within panels.*

DECORATIVE WALLS 175

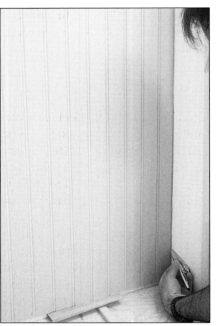

Figure 11-34. Start with the first piece in the corner, shimming it so that the grooves are plumb, and then scribe it to fit the corner.

Figure 11-36. To be sure the corner piece will be parallel (and to determine how much to spread your scribes), measure the distance from the edge of the board to the mark on the top of the preceding piece. Transfer that measurement to the bottom of the piece.

Figure 11-35. To scribe the last corner piece, first mark its position on the top of the preceding piece.

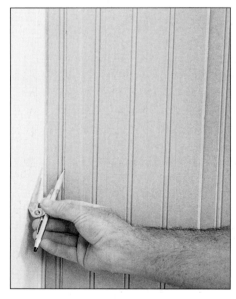

Figure 11-37. Spread your scribes that exact measurement, position the corner piece on the two marks, and then scribe the wall contour onto the corner piece.

muster and gets installed. Personally, I believe that narrow recessed panels add drama to a room, but before I install any material and create a melodrama, I have the GC and the owner approve the layout.

Crown Details

Another area to consider while laying out full-height paneling is the ceiling. The top rail must be sized to accommodate both crown molding and ceiling paneling. If the stiles are 4 in. wide, I try to allow a 4-in. reveal between the crown molding and the top of the panel molding. Opinions vary, but I believe that anything short of a two-thirds exposure is deficient—a reveal much less than 3 in. wouldn't look nearly as good.

Cabinet Details

I treat cabinets in much the same way I do other openings—I box around them with stile material, leaving the same reveal as I do around doors and windows. But rules are made to be broken: I dislike the box-out around the upper bar cabinet shown in Figure 11-31. That wall unit hangs separately from the base unit, and the paneled wall would have been improved if I had terminated the panel molding right into the cabinet.

Solid-Sheet and T&G Paneling

For beadboard and sheet paneling, I follow many of the same techniques I use for raised paneling; however I snap my line at the top of the paneling rather than at the top of the chair rail.

But scribing the first piece to the wall is the most critical part of the job. If the first sheet doesn't follow the snap line and fit snugly against the corner wall, then by the second piece the job can turn into a nightmare. Even a slight discrepancy can be compounded enormously by the third piece, especially if the material comes in 4-ft. sheets!

I cut all the pieces so that they're about a half-inch short of the floor, so that I won't have to fight any humps in the floor — the baseboard will later cover any gaps. For sheet goods, I use my table saw or a circular saw and cut-off guide.

I temporarily block the first piece off the floor, hold the top perfectly flush to the line, and scribe the end to the corner wall, just as I do the first stile for recessed paneling *(Figure 11-34)*.

The field pieces are easy to install, but the final corner piece must be scribed to the wall, too. The backward-thinking technique is exactly the same as scribing casing. First, block the new piece up and place it tightly against the corner. Draw a small pencil line on the top of the preceding piece, at the back edge of the new piece *(Figure 11-35)*. To ensure that the new piece is parallel with the edge of the preceding piece, measure the distance from the edge of the preceding piece to the pencil line, and then transfer that mark to the bottom of the preceding piece *(Figure 11-36)*. Note: whether the material comes in 4-ft. sheets or 4-in. boards, the technique is exactly the same.

Spread your scribes that exact measurement, hold the new piece firmly against the preceding piece so that it's aligned with the pencil marks, and trace the contour of the wall onto the leading edge of the new piece *(Figure 11-37)*. The first cut is nearly always right. (See "More on Scribing Techniques," next page.)

CHAIR RAIL

Like many aspects of trim carpentry, chair rail can be designed in countless ways. The example in Figure 11-11 is a simple "reversible" pattern, easy to use as a typical plant-on molding. But chair rail can be assembled from several combined moldings with dramatic results, regardless of the casing thickness.

The thickness of the casing often creates a stumbling block for carpenters designing wainscoting because the chair rail is always proud of the paneling (Figure 11-12), and frequently proud of the casing. Many carpenters attempt to avoid this apparent problem, but the troublesome detail can be used for a dramatic effect.

Chair rail designs in historic homes often provide answers to design dilemmas. I found the detail in *Figure 11-38*, which resembles a small mantelpiece, at a Georgian-style home in Tennessee, though I made the top cap of MDF. I ripped the material and

Figure 11-38. *An attractive two-piece chair rail can overlap the casing, but best if only marginally.*

Figure 11-39. *Cut the top cap long, and then scribe the horn so that it overlaps the casing (left). For tighter stain-grade joints, scribe the cap twice — once for a rough fit, and once more for a final fit (right).*

> ## More On Scribing Techniques
>
> Even if you're scribing paneling to the multiple moldings on a mantelpiece, follow exactly the same routine. Remember, rather than spreading your scribes the distance between the paneling and the mantelpiece, measure and spread your scribes the amount that the back edge of the paneling projects past the leading edge of the previous piece *(see photo)*.
>
> Don't be fooled by an intricate mantelpiece or daunted by a stone wall. Just follow the profile carefully with your scribes and then cut slowly with a jigsaw and fine-tooth blade. No matter how elaborate the scribe, if your first attempt is close (as it often is), the second pass can be pleasingly perfect.
>
>
>
> Even an elaborate mantelpiece leg yields to careful scribing and cutting.

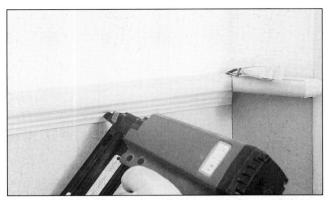

Figure 11-41. *Glue up the self-return ends first, and then attach the longer field pieces while the glue is drying.*

routed both edges with a 1/4-in. roundover bit. I adjusted the depth of the router so that the bit would also cut a slight step, which adds a little punch to the bullnose. I ripped the cap wide enough to allow for the bullnose, the step, and a very small reveal between the molding beneath and the cap above.

The length of the horn or ear that projects past the casing shouldn't be so long that it covers the profile of the casing—the horn shouldn't interfere with the casing. Besides, scribing the horn is much easier if it terminates before the casing profile begins. In this example, I cut the horn long enough to fill most of the initial flat step in the casing.

For a tighter stain-grade fit, expect to scribe and cut the piece twice: first for a rough fit, second for a finished joint. For both attempts, spread your scribes exactly the distance from the wall to the back of the piece *(Figure 10-39)*. Use a jigsaw or coping saw to make the cuts, but don't back-cut — hold the blade perpendicular to the material — or the bottom of the cap won't fit tightly against the casing.

A variety of different moldings can be applied beneath the top cap. Because I had an ample supply of reversible chair rail left over from another job, I ripped that molding in half. The classical-ogee profile also matched the recessed panel molding and added more detail to the chair rail, without overpowering the room.

Always try to cut self-return pieces early on, to allow time for the glue to dry before installation. The length of the self-return should allow for the same reveal between the top of the molding and the bottom of the cap. A small spring-loaded hairclip works well for temporarily securing the glued-up corner *(Figure 11-40)*.

DIAGONAL PANELING

Up to this point, all the paneling types that I've discussed share many of the same installation techniques — install one and you can pretty much install the other. But diagonal paneling, especially with an outside corner, is a different kind of animal: It's big, hairy, growls a lot, and bites hard, but once tamed, it'll be your friend for life.

The angle of the molding on the wall in *Figure 11-41* was determined by the slate fireplace surround. I started out by finding the angle at the wall on both sides of the corner. A bevel square helped

Figure 11-41. *Diagonal paneling is a challenge but the result can be dramatic, especially on sharp outside corners.*

Figure 11-42. *Even with compound miters on both ends of each piece, proper layout and preparation will allow the job to proceed efficiently.*

me to get started, but trial and error yielded the exact miter and bevel settings for my compound miter saw. The corner was tricky, though, because the ceiling panels (installed earlier by my finish crew) followed the waves in the framing and weren't perfectly straight or flat. I knew that each piece would have to be scribed into the corner in order to achieve tight-fitting joints.

I approached the acute angle on the outside corner in much the same way. First I cut a few test pieces and determined the exact miter/bevel settings for both sides of the wall — they had to be identical or the paneling wouldn't climb the wall evenly.

While holding two test pieces against the wall, I found that the short point of the paneling had to be 1/4 in. past the corner bead in order for the long point of the paneled corner to align flush with the sharp slate corner. After that discovery, I checked to see in which direction the corner was out of plumb. Before installing the first piece, I had to know if the corner leaned out or in at the top. If it leaned out, I'd have to remove the corner bead and maybe some of the drywall, too, before I started installing any material.

I found that the corner did lean out, but only about 3/16 in., which fell within the 1/4-in. gap I'd have at the bottom. With that issue resolved, I faced one last question: I debated for several minutes whether or not I should install the material with the tongue up or the groove up. After trying a few scraps, I realized that if I started with the groove up, I'd have to shave a little off the long point of the tongue in order to slide each board into place. But if I started with the tongue up, I'd have to shave much more off the back of each board so that each would fit over the long-point of the corner. I decided to run the material with the groove up.

Now I was ready to start cutting boards. To minimize the number of times I'd have to change miter/bevel settings, I cut a good stack of pieces—

in varying lengths and allowing plenty of extras for mistakes — for each side of the wall. I cut both the compound miter for the outside corner, and the compound miter for the inside corner on all pieces. I knew that the settings for the outside corner would remain the same all the way up the wall, but I couldn't trust the inside corner. With the inside corner cut on every piece, I'd be able to hold the entire board in position and judge the joint or make the scribe as necessary.

All the preparation time paid off. I was able to install three or four pieces on one side of the wall before turning to the opposite side, where I'd catch up and then run past by another three or four boards *(Figure 11-42)*. Before long, I reached the ceiling. After several attempts, I finally cut the last little triangular piece, with miters carefully back-filed by hand, so that it just fit. My only regret was that I became too absorbed with the work and failed to take enough photographs.

Chapter 12

Decorative Ceilings

I've always believed that man wasn't meant to stand on scaffolding that's more than 4 or 5 feet off the ground. But these days, with both the amount of ceiling work and the size of houses on the increase, I've been working at dizzying heights. Even these elaborate projects are bid competitively, however, and profits aren't keeping pace with the price of scaffolding. Competition always forces carpenters to think of innovative ways to increase productivity and still ensure quality.

While some of the methods described in this chapter might seem like shortcuts to a few seasoned carpenters, keep in mind that composition molding was once considered a shortcut to hand-carved woodwork. Installation procedures always change as carpenters adapt to new products. Some of the techniques here are necessary responses to technical improvements in materials, adhesives, and fastening systems.

Decorative ceilings vary in design, from true coffered ceilings with a network of beams and crown molding, to simple flat paneling made from a grid of stiles, rails, and panels, similar to the wainscoting and library paneling covered in Chapter 11. Somewhere in between are light wells and decorative soffits.

PANELED CEILINGS

This chapter follows on the heels of the last chapter because wall paneling and ceiling paneling share many of the same features: the fastening and installation procedures are mostly the same, and the sizes of the stiles, rails, and panels are similar. However, laying out a ceiling is usually much more demanding than laying out wainscoting or library paneling.

Layout

Panel sizes vary dramatically from one style of home to the next. Some rooms are designed with small, equally sized, square panels. Unless the ceiling is extremely high, this arrangement decreases the apparent size of the room, which is often the intention in a library or den — a smaller space evokes more intimacy. In a great room or living room, long rectangular panels tend to increase the perceived size of the space, especially along the length of the panels. A mixture of panel sizes is also popular, with small square panels in the corners and larger rectangles in the field. In some cases, abbreviated perimeter panels are meant to suggest the continuation of the field panels beyond the confines of the room's walls, which also adds to the apparent size of the room.

With all these choices, clients often have a difficult time choosing a paneling design. Preferably, paneling designs should be approved on paper before subcontractors install flush lights, light fixtures, heat-and-air-conditioning ducts, stereo speakers, fire sprinklers, and alarm sensors. In that case, my work begins as soon as the room is framed. I lay out the ceiling design on the rough floor and snap chalk lines for all stiles and rails, then spray clear sealer over the lines so that they'll still be there after the drywall crew scrapes and washes the floor. With the ceiling design snapped on the floor, all the subcontractors use plumb-bobs, lasers, or string lines to place their utilities.

When I work on ceilings where the drywall has already been installed, I scratch my head a lot more — for two reasons. First, laying out these ceilings becomes a wrestling match between design and cost, so I spend more time thinking. Second, it's better to move a few can lights or an occasional heating duct than to leave large items off-center in a panel, so I spend more time brushing bits of drywall out of my hair. Usually, utilities can be moved without too much difficulty, especially since the paneling will cover the entire ceiling.

Whether I'm working on the floor in a new house or on the ceiling in a remodel, I always begin by measuring and snapping chalk lines around the perimeter of the room. The perimeter rails must be wide enough to accommodate any crown molding at the corner of the wall and ceiling, and still leave enough reveal to match or resemble the stiles in the field of the ceiling. A minimum $2^{1}/_{2}$-in. reveal is required — beyond the crown molding — if the field stiles are 4 in. to 5 in. wide.

Figure 12-1. *Lay out and install the perimeter rails first. Use glue, splines, nails, and adhesive to secure the material and the joinery.*

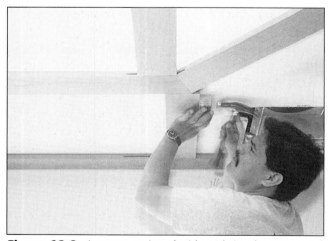

Figure 12-2. *A router equipped with a $^{1}/_{4}$-in. slot cutter and MDF splines produce perfectly registered joints that require little sanding. Cut the field stiles tight and slip them in between the rails, using a pry bar and a small block of softwood — not brute force — to coerce the pieces into place.*

Once the perimeter is marked, I start laying out the panels in the field. I like to snap out the entire design and win approval before starting the actual woodwork. I'd rather spray out the old lines with fast-drying primer and snap a new layout on the ceiling than build the ceiling twice. To ensure that the client can completely visualize the design, and to make installation quick and easy, I snap both sides of each stile.

Installing the Stiles and Rails

On ceilings, I install the perimeter rails first **(Figure 12-1)**. Whether the ceiling is paint-grade or stain-grade, I use a $^{1}/_{4}$-in. slot cutter in my router and MDF splines to join all the mitered corners and field splices (see Figure 12-10, and Figure 11-13, page 168). Using carpenter's glue, glue in all the joints, along with paneling adhesive between the rails and stiles and the ceiling, ensures that the pieces won't move and the joints won't open. Wherever possible, I nail the framing components into the ceiling joists.

I miter all the corners in the perimeter rails, which is also why they must be installed first. Afterward, I cut the interior stiles to fit snugly between the rails. I slide the stiles in from the side, and use a pry bar to ease each piece over the spline. Because the pieces are cut to fit tightly, a tap or two against a softwood block is often required **(Figure 12-2)**.

Oddly Shaped Panels and Panel Molding

I use the same techniques for cutting the panels and the panel molding as those discussed in the previous chapter. However, on a ceiling I often encounter strange-looking panels, especially L-shapes and acute angles that are sharper than 60 degrees. Measuring and cutting panels and panel molding for these unusual shapes can be challenging, but a few simple techniques help to speed up the process considerably.

L-shaped panels. When measuring L-shaped panels, carpenters often subtract too much for wiggle room; no one should lose faith in hard-learned measurement and cutting skills. Start by measuring the four legs that frame the outside corners. Subtract $^{1}/_{8}$ in. from each of those measurements. Next, measure the two legs that form the inside corner, but don't subtract anything from those measurements **(Figure 12-3)**. The finished panel

Figure 12-3. *Measure L-shaped panels carefully: Start by measuring the four legs that frame the outside corners, and subtract 1/8 in. from each of those measurements. Next, measure the two legs that form the inside corner, but don't subtract anything from those measurements.*

Figure 12-4. *Measure and cut carefully, and even the strangest shaped panels will fit perfectly the first time. Use plenty of panel adhesive to secure the panel to the ceiling.*

should end up 1/16 in. short all the way around. For that reason, always subtract 1/16 in. when measuring for utility cutouts.

I always draw a diagram of oddly shaped panels in the small notebook that I carry in my tool bag — especially if the panels have cutouts for lights, ducts, and such. I draw the diagram using the lines on the notepaper to denote the direction of the grain — that way I never have to worry about which way the grain runs.

Before transferring the measurements from paper to panel, always turn the panel upside down so that the panel won't be a mirrored image. Also, by cutting from the back of the material, the saw won't tear out the face grain. Use these simple techniques, and even the most difficult panels will always fit just right. Without question, the panel doesn't have to fit that tightly because the molding will cover nearly an inch of the inside perimeter. But to guarantee that all cutouts are positioned properly, the panel must be measured and cut accurately. If you plan to attach the panel to the panel molding before installing the panel, a perfect fit is a must (see Chapter 11, Figure 11-27). Either way, apply a liberal quantity of panel adhesive to the back of the panel before installation *(Figure 12-4)*.

Cutting acute-angle panel molding. That same measurement technique — in fact the same rough drawing — works well for cutting the molding around L-shaped panels, too *(Figure 12-5)*. By using the same measurements, the panel molding and the panel will be perfectly matched. For cutting

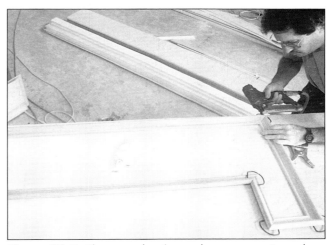

Figure 12-5. *The same drawing and measurements work perfectly for cutting the panel molding. To ensure tight-fitting miters, preassemble the molding. Nail through the back of the miters so the joints won't open.*

panel molding, I use the cutting jig mentioned in the previous chapter (see Chapter 11, Figures 11-21 and 11-22). However, the molding around some panels requires a little extra attention.

As I mentioned in Chapter 7 (see Figure 7-29), the miter scale on a chop saw — except for the 45-degree mark — doesn't represent the true angle of the cut. Because the "0" mark is actually 90 degrees, the "15" mark is really 75 degrees, the "22 1/2" mark is 67 1/2 degrees, the "30" mark is 60 degrees, and the "60" mark is, in fact, 30 degrees.

Even a chop saw with marks up to "60" is

DECORATIVE CEILINGS 183

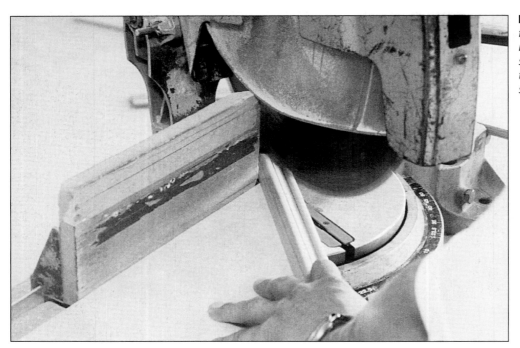

Figure 12-6. *An accessory fence is required for cutting miters at corners that are sharper than 60 degrees — the maximum that most chop saws can miter.*

Figure 12-7. *Be certain to preassemble acute-angle panel molding because the miters will never fit precisely if the pieces are installed individually.*

Figure 12-8. *Acute-angle panel molding doesn't always fit perfectly the first time, so install the panel first, then the molding. To avoid unnecessarily large nail holes, secure the panel molding to the stiles and rails with 18-ga. brads.*

capable of cutting miters for angles only between 90 and 60 degrees.

To cut miters for angles that are more acute than 60 degrees, I use an accessory fence. I cut a wide board at a 45-degree angle and clamp the board to the extension wing on my miter saw *(Figure 12-6)*. Since the accessory fence is cut at 45 degrees, if the saw is set at the "0" mark, the cut will be exactly 45 degrees. If the saw is moved from the "0" mark, the amount it is moved should be subtracted from 45 degrees. For a 40-degree corner requiring 20-degree miters, I swing the saw to the "25" mark on the miter-saw scale: 45 minus 25 equals 20.

Figure 12-9. *Case the bottom of light wells first, before installing the vertical trim or glass stop. Calculate (or guess at) the degree of the miter, and then cut test pieces for each corner. Note the exact miter at each corner before cutting any pieces to exact length.*

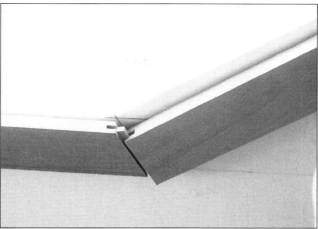

Figure 12-10. *Splines are better than biscuits on miters, too, and guarantee flush joints.*

Figure 12-11. *The center stile is slightly wider than the drywall covering the beam — about $1/16$ in. proud, so that the drywall won't interfere when it comes time to install the vertical trim. Splines secure all the joints.*

Installing the panels and molding. Acute-angle panel molding should always be preassembled; otherwise, the miters will never be tight *(Figure 12-7)*. As with wainscoting panels, I'm inclined to attach the ceiling panels to the back of the panel molding before installing either, but because I don't cut that many acute-angle panels, I'm rarely confident that they'll fit perfectly, so I install the panel first and then the molding *(Figure 12-8)*.

LIGHT WELLS AND SOFFITS

Like the perimeter of a room, the flat molding on the ceiling around a light well must be installed first. Think of the molding on the ceiling as the jamb, while the molding inside the light well is similar to the casing. Of course, some designs are completely opposite: occasionally the ceiling molding laps over the light-well molding, in which case the ceiling molding would be installed last. I find that mocking up the pieces before installation is the best method for ensuring mistake-proof productivity.

Simple Moldings

I worked on the ceiling pictured in *Figure 12-9* with one of our lead finish carpenters (mostly I shot photographs while he applied the molding). To find the miter angles on this octagon, we divided 360 degrees by 8, and then divided the result, 45, in half. Framing from calculations is never perfect, so we cut two pieces of scrap material at $22^{1}/_{2}$-degree angles and tried them in each corner. Then we noted the precise miter needed for each pair before cutting any pieces to exact length.

We installed the perimeter pieces $1/16$ in. proud of the light-well soffit so that the well walls wouldn't interfere when it came time to apply the glass stop (more on the glass stop in a moment). We secured the miters on the perimeter rails with splines *(Figure 12-10)*. Next, we ripped the two center stiles slightly wider than the crossed beams they cover — again, so the glass stop could be applied without interference. Like the stiles on the ceiling, we cut tight-fitting butt joints and secured them with splines *(Figure 12-11)*.

The glass stop was installed last and covers the MDF edge of the rails and stiles *(Figure 12-12)*.

Figure 12-12. *To ensure safety and accuracy, we use an accessory fence to cut the glass stop at angles that are more acute than 60 degrees, and then carefully tap those long sharp corners into the opening.*

Figure 12-15. *Backing was added to the top of the light-well beam to hide the electrical conduit.*

Figure 12-13. *We installed the pieces one-at-a-time; alternatively, preassembling all the pieces for each opening might have saved us time and resulted in tighter-fitting miters.*

Figure 12-16. *Rounded inside corners are troublesome for finish carpenters because the drywall must be cut carefully out of the way of the miter.*

Figure 12-14. *Careful layout and simple production techniques can produce richly paneled rooms in a fraction of the time (and at a fraction of the cost) of traditional methods.*

Using an acute-angle accessory fence made cutting several of the angles a simple task. We chose to install each piece individually, rather than risk pre-assembling the entire frame. To keep the nail holes less visible, we used a brad nailer and glued all joints to ensure that they wouldn't open *(Figure 12-13)*. I'm convinced we could have improved the joinery slightly and shaved the installation time if we had preassembled all the pieces. Nevertheless, it came out very well *(Figure 12-14)*.

Combination Moldings

Crown molding, baseboard, and casing are often enriched by a combination of moldings. Stacking different moldings is also an attractive way of adding detail and drama to light wells (see "Flex-

Figure 12-17. *Always install the shortest pieces first, and then bow the longer piece into place.*

Trim Corners"). Besides, the strong raking light from a light well is ideal for accentuating both subtle and complex molding profiles. Once again, molding combinations follow the precedent of classical architecture because the details in a deep light well resemble a classical entablature.

As I mentioned above, sometimes the flat ceiling molding is installed first, and sometimes the vertical light-well molding is installed first — it depends on which molding overlaps the other. Regardless of which piece must be installed first, working from a ladder or scaffold, it's most productive to complete all the trim in each well opening before beginning a new opening.

In the light well shown in *Figure 12-15*, we added additional backing to the top of the beams to hide the electrical conduit leading to the light fixture. Then we laid out the molding on the vertical faces of the well. Of course we had to carve out the drywall corners next. Drywall installers tape sharp 90-degree inside corners, but in my area any corner greater than 90 degrees is floated with a radius tool. Those corners are a nightmare for trim carpenters, because the long point of the miter must be buried in the drywall *(Figure 12-16)*.

The lower trim, or cap, is applied first. We butted the narrow edge of the cap against the narrow edge of the casing on the ceiling. Because the material was paint-grade, all the joints were caulked with adhesive latex. The short pieces were installed first because the longer pieces could be bowed into place *(Figure 12-17)*.

The upper trim is exactly the same molding but installed in reverse, with the narrow edge up *(Figure 12-18)*. To finish the square edge of both

Flex-Trim Corners

Trimming out a light well with radius corners is another example of how useful flex trim can be for paint-grade or stain-grade ceilings. Paint-grade flex trim is smooth, while stain-grade flex trim is embossed with faint wood graining and can be faux-finished to resemble natural wood.

But no matter which material is used, the installation technique is always the same. Temporarily tack all the pieces to the ceiling, with the joints overlapping in each corner *(Figure A)*. Next, use a square and transfer the long-point marks carefully across both pieces in each corner *(Figure B)*. Mark the short points, too, but go slowly — if the marks are made perfectly, and each piece is cut precisely, the joints will usually fit just right the first time — especially for paint-grade work.

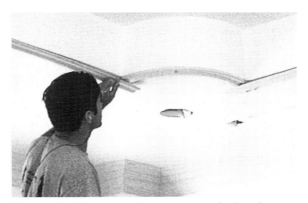

Figure A. *Miters on radius corners — whether the material is flex trim or wood — are easier and faster to cut precisely if the material is first tacked in place.*

Figure B. *Once the pieces are temporarily secured, mark both the outside long points and the inside short points of the miters on each piece. Use a square to be sure the lines are perfectly plumb and straight.*

Building the Backing

Finger-joint jamb stock is the most economical and trustworthy material to use for building the backing boxes. Avoid MDF because nails into the edge split that material too easily. A good length for these U-shaped boxes is 12 in., but the width of the backing box must be exactly the width of the finished beam. Rip the inside top plate down to allow for the thickness of the two sides. I prefer to shoot nails through the sides and into the edge of the inside top plate because it's stronger than nailing down through the top plate into the tops of the sides. The width of the sides should equal the depth of the finished beam, of course minus the thickness of the beam bottom, which is applied to the edge of the sides.

If the beams are all flush, build special T-shaped backing for the beam intersections. To end up with 12 in. of solid backing for each beam (on either side of the intersection), the top plate should be about 24 in. long, plus the width of the beam. Place two shorter 12-in. pieces of inside top plate perpendicular to and at the center of the longer piece *(Figure A)*. Then glue and nail another 12-in. piece that spans the intersection *(Figure B)*. Finally, attach the sides of the intersection. Because the sides determine the fit of the baseboard and crown joinery, take special care to fasten the sides so that the inside corners are flush and the sides perfectly square *(Figure C)*.

Figure A. *Build special backing for all intersections. Start with a long piece of inside top plate, and then place two 12-in. pieces of top plate at the center and perpendicular to the longer piece.*

Figure B. *Secure the three-piece corner by overlapping, gluing, and nailing a fourth piece.*

Figure C. *Attach the sides so that they're flush on the inside corners; this will ensure perfectly square and plumb backing for the baseboard and the crown molding.*

pieces of trim, we installed a rabbeted panel molding, similar to a back band. The panel molding was also installed in reverse: the narrow edge on the lower piece points up, while the narrow edge on the upper piece points down. Because the trim pieces are the same on both the top and the bottom, the entire pattern is called "reversible" and resembles an oversized, classical, astragal molding *(Figure 12-19)*.

COFFERED CEILINGS

Typically, coffered ceilings involve a lot of work: layout and backing during rough framing; layout and installation of additional backing after drywall; the painstaking assembly of false beams across a ceiling; and finally, several stages of built-up moldings within each coffered opening. But a stunning ceiling doesn't have to mean a stunning price tag. By using a short-cut backing system, cof-

Figure 12-18. *All the molding is identical, though the upper pieces are installed upside down.*

Figure 12-19. *From a distance, the combination of moldings resembles a classically inspired "reversible" astragal.*

fered ceilings can be installed at competitive prices in even modest homes.

A Hollow-Backing System

For backing, we build hollow U-shaped boxes as in *Figure 12-20* (see "Building the Backing"). The boxes, each about 16 in. long in the field and 8 in. to 12 in. long around the perimeter, are fastened to the ceiling with nails and adhesive caulk. Next, the continuous bottom cap or "beam bottom" is applied to the bottoms of the U-shaped boxes, spanning from box to box. Then the base molding is installed, nailing into the edge of the cap, also spanning from box to box. Finally, the crown molding is installed, which is nailed into the top of the base molding, into the U-shaped boxes, and into the ceiling — especially in the ceiling joists. But the success and speed of the process depends on precise layout.

Layout, layout, layout. I always start a coffered ceiling by nailing together a mockup of the finished beam and crown molding. That way I know exactly how wide and how deep the beams will be. I usually lay out these ceilings as soon as the room is framed, before the installation of any electrical, heating, or other components. As with paneled ceilings, I measure and snap chalk lines on the floor after framing. Tradespeople then plumb up from those lines, as indicated earlier. I note the exact measurements of the layout on my plans, just in case the lines aren't visible after the dry-

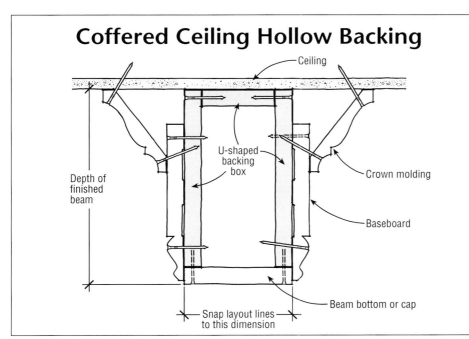

Figure 12-20. *Hollow U-shaped boxes 8 in. to 16 in. long provide backing for false beams. The boxes are fastened to the ceiling with nails and adhesive caulk and are tied together by a continuous bottom cap spanning from box to box.*

DECORATIVE CEILINGS

Figure 12-21. *Careful and precise layout is the first step for installing all ceiling work. Snap chalk lines for BOTH sides of each beam.*

Figure 12-22. *The backing is constructed hollow so that a nail gun fits easily between the pieces. But don't rely on nails alone; use plenty of panel adhesive, too.*

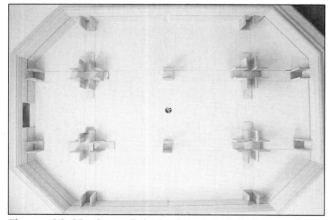

Figure 12-23. *Once all the backing is installed, the ceiling looks like a geometric puzzle, but don't hang any trim until all the adhesive has had at least one night to dry.*

wallers wash the floors.

After the drywall is finished, I snap lines for both sides of each beam, exactly the width of the hollow box or beam bottom *(Figure 12-21)*. In some cases, I install a half-beam around the perimeter of the room — in which case, I snap a line for that backing, too, at a distance from the wall equal to the width of the half beam. In rooms without half-beams on the wall, I snap lines for the bottom of the baseboard molding.

Installing the backing. The backing boxes should be installed approximately 24 in. on-center, with an additional box against the wall at the end of each beam. Because the U-shaped boxes are hollow, it's easy to get a nail gun inside for fastening, but attachment depends primarily on panel adhesive, so use a liberal amount. I use Liquid Nails® (see Appendix 1), especially where the backing misses a ceiling joist *(Figure 12-22)*.

On the first day of a ceiling installation, it's best to schedule enough help to lay out the ceiling, snap the lines, build the backing boxes, and fasten them to the ceiling. Allow the adhesive to dry overnight before applying the beam bottoms *(Figure 12-23)*.

The beam bottom is easy. I use MDF material for beam caps, whether I'm installing a paint-grade or a stain-grade ceiling. For paint-grade ceilings, I use 16-ft. lengths of MDF wall-cap for the bottoms of the beams. From my suppliers, I'm able to buy wall-cap in a variety of widths from $3^{1}/_{2}$ in. up to $7^{1}/_{2}$ in. For stain-grade ceilings with longer spans, I order custom lengths of veneered MDF-core sheet goods in 10-ft. or 12-ft. sizes. I've found that MDF is more stable, is more uniform in width and thickness, and finishes more evenly than solid wood. Sheet material is also far less expensive than lumber.

Whether you're working at exceptional heights off a scaffold or on a normal ceiling off a ladder, the beam bottoms are easy to install. Start with the perimeter beams *(Figure 12-24)*. Next, fasten the bottoms to the main beams *(Figure 12-25)*. If the ceiling isn't flat, stretch a string from one wall to the other and drive shims between the beam bottom and the hollow backing. On a high ceiling, the beams don't have to be perfectly straight, but large dips and dives on an 8-ft. or 10-ft. ceiling can be visible, so take the time to keep the boards straight.

All field joints can be butt cuts. They should be tight, but the joints don't have to be perfectly flush. Just be sure that every joint falls at an intersection so that they'll be covered by bosses, rosettes, or pendants (see "Bosses, Rosettes, and Pendants," page 192).

Figure 12-24. *First, install the bottoms on the perimeter half-beams. For perfectly flush joints, use a router, 1/4-in. spline cutter, and 1/4-in. MDF splines.*

Figure 12-25. *Next install the beam bottoms in the field. All the joints must land at the intersections, so that they'll be covered later by rosettes, bosses, or pendants.*

Preassemble the beam sides. The sides of the beam are made with tall baseboard, installed upside down. Almost any profile will work, though for fastening the crown molding, it's best to have a wide flat area. The wider the flat area, the less chance that dips and dives in the ceiling will show up in the reveal between the bottom of the crown molding and the first detail on the baseboard.

The profile on the baseboard is also important because it determines the appearance of the beam. A simple ogee profile, or single bead and quirk, help to simulate the look of a simple flat beam with a single dressed edge; a more elaborate profile lends more detail and depth to the crown molding.

In either case, I've found that preassembling the baseboard has several advantages: First, the joints can all be cut with simple miters, and then glued and nailed through the back. Second, if the pieces are preassembled there's less risk of pushing the backing out of position. And finally, it's simply faster to assemble the pieces on a large comfortable workbench than while standing on a ladder or scaffold. To speed up the process and ensure tighter joinery, scaffold and ladder work should be limited to nailing the boxes in place *(Figure 12-26)*.

The crowning touch. Preassembly is also the *only* way to fly when installing the crown molding — and the job literally flies along at this point

Bosses, Rosettes, and Pendants

Applying ornamental trim to the beam intersections is the final touch for most coffered ceilings. These rosettes (or bosses) are easy to make and offer finish carpenters a chance to piece together some original and attractive designs. We assembled a two-piece boss for the high classical ceiling in **Figure A**. By fastening the smaller rosette into the center of the larger assembly, we increased the depth and drama of the profile. Such a large rosette would be ostentatious on a normal ceiling **(Figure B)**.

In most cases, baseboard or casing can be cut to make the rosettes. To ensure that the rosette will cover the joints, cut the molding so that the long points exceed the width of the beams by 1/16 in. to 1/8 in. **(Figure C)**.

Pendant bosses for Victorian-style homes can be made from crown molding. This is a rare chance for a finish carpenter to fit together all those cutoffs and turn scraps into stunning ornaments **(Figure D)**.

Figure A. *The dessert in a coffered ceiling is making the rosettes or bosses for the beam intersections. These mitered squares are easy to cut and preassemble from baseboard or casing.*

Figure C. *For lower ceilings, cut small rosettes from the same casing or baseboard used throughout the house.*

Figure B. *For a high ceiling, combine several layers of molding to increase the depth of the rosette. Deep shadow lines will be more visible from the floor.*

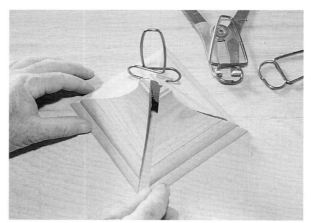

Figure D. *Victorian and Arts and Crafts–style homes are often decorated with carved pendants, but scraps of crown molding — mitered at the corners — make a handsome site-built ornament.*

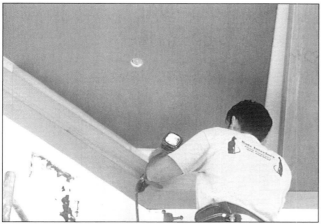

Figure 12-26. *Preassemble the baseboard pieces for each coffer so that the joints can be glued and nailed from behind.*

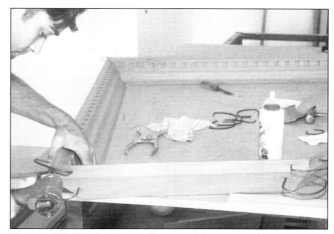

Figure 12-28. *Precise and successful preassembly goes easier with strong miter clamps. Use the clamps to secure the joint while shooting fasteners through the miter.*

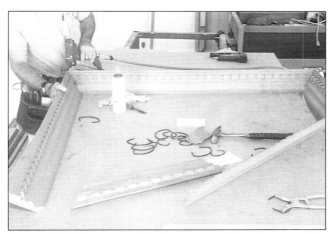

Figure 12-27. *Preassemble the crown molding, too, and the joints will not open or separate.*

Figure 12-29. *A two-man crew can hoist awkward panels up to high ceilings with a long rope.*

(Figure 12-27). Of course, if the ceiling design calls out for solid panels, be sure to install the panels before the crown molding. (Tight-fitting panels are obviously not required; the crown molding covers more than 2 in. of the panel's perimeter, but secure the panels well with adhesive and brads.)

Measure the crown molding just like the baseboard — precisely. Cut the molding with the same care, and use miter clamps to hold the corners tightly in position while nailing the joints from behind *(Figure 12-28)*.

To save time moving a ladder or scaffold around the room, cut and install each of the squares one at a time — both the baseboard and the crown molding. That practice will also prevent any over-sized molding squares from shoving the backing a little out of position.

Use a rope to pull the assembled molding up to the top of the scaffolding *(Figure 12-29)*. Even out-of-square coffers should be preassembled because it's the only way to ensure tight-fitting joints without splaying the backing. The assembled squares should fit snugly into the coffers, driving air out as they're pressed into place *(Figure 12-30)*. Because of their small, almost invisible heads — yet more than adequate length — brads are the best choice for securing the molding to the beam *(Figure 12-31)*.

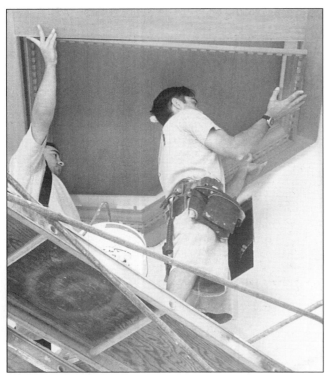

Figure 12-30. *The panels, especially the crown molding, should just fit into the opening — forcing out a whoosh of air and a sigh of relief.*

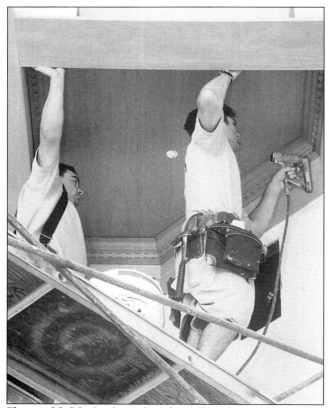

Figure 12-31. *Brads are best for nailing the molding to the backing, but use 2½-in. finish nails to secure the crown to the ceiling, and nail into every joist possible.*

Figure 12-32. *Begin working a radius corner from a line measured down from the ceiling and representing the bottom edge of the baseboard. Determining the length of flexible molding along that bottom line can be difficult with a tape measure, but a scrap piece of flex-trim works well and takes the hassle out of the job.*

Figure 12-33. *The flexible baseboard is fastened to the wall.*

Figure 12-34. *The remainder of the baseboard can be coped to fit the radius baseboard, while the other corners can be coped or mitered.*

Figure 12-35. A chalk line doesn't work too well on a radius wall, though a pair of scribes does, to ensure that the crown molding is positioned for a straight and even reveal between it and the nearest profile on the baseboard.

Figure 12-36. The straight crown is then coped to the radius crown and installed.

But use 2 1/2-in. nails to fasten the crown to the ceiling, and try to locate every joist possible so that the crown molding will also help secure the entire decorative diaphragm.

Radius Coffers

Radius corners in coffered ceilings *(Figures 12-32 to 12-36)* aren't rare anymore, perhaps because radius trim isn't nearly as expensive as it once was. Flex trim has reduced both the price of the material and the necessary skill to install paint-grade radius work. Don't get me wrong, I don't mean any disrespect for the skills of my fellow carpenters, but milling radius trim from stock lumber doesn't fall within the limits of job-site finish carpentry. Besides, on most jobs we don't have room in our trucks—for all the tools—or in the budget to afford solid-wood radius trim.

SOLVING CEILING PROBLEMS

This won't come as a surprise to experienced finish carpenters, but beginners should always expect that framing, rough openings, and structural layout will never be correct. Sometimes it seems that the primary job of a finish carpenter is to solve problems. Certainly the primary responsibility of a finish carpenter is to make everything look good, preferably before the painter arrives.

Fortunately, a lot of corrective work can be accomplished on a decorative ceiling. Even if a light well, soffit, or an octagonal recess *(Figure 12-37)* is framed incorrectly, a decorative ceiling can provide the opportunity for corrective cosmetic surgery.

Before I laid out this ceiling I recognized that the soffit depths and reveal lines between the cabinets and the recessed octagon varied considerably. Had I not corrected that glaring error, the finished ceiling would have looked terrible, no matter how nicely I installed the new trim.

To correct this ceiling, we came up with a design that covered all the drywall. New casing would cover the ceiling, right to the cabinets; a course of inverted baseboard would cover the vertical face of the octagon; and the coffered ceiling

Figure 12-37. The reveal between the crown molding and the cabinets varied, especially across the short corner of the octagon. But the trim design for the new ceiling would cover every inch of the recess, as well as the little bit of ceiling up to the cabinets (see mock-up in upper-right corner of photograph).

Figure 12-38. *I cut MDF backing for the corners and beveled both ends of each board. The ends would soon be covered by trim, but the bevels made it easier to locate and install the backing with precision.*

Figure 12-40. *The last step in this "fix" was to install the inverted baseboard, which butted up against the casing on the ceiling.*

Figure 12-39. *After filling in the corners so that the soffit was consistent with the cabinets, I was able to rip the casing for the ceiling to 2⅞ in., then installed it flush with the face of the recess.*

would cover the remainder of the recessed opening. So the first thing I did was to remove the old crown molding from around the cabinets. After installing MDF backing in each corner, I started applying the moldings. It wasn't long before the ceiling looked just right *(Figure 12-38 to 12-40)*.

CHAPTER 13

Decorative Doorways

Doorways, more than any other detail, reveal both a home's architectural style and the relationship of that style to classical architecture *(Figure 13-1)*. The precedent of classical styles and the current fashion of mixing classical styles are good reasons for carpenters to brush up on their knowledge of classical architecture. But there's more that carpenters need to know about their craft. Many Georgian, Federal, Victorian, and Arts and Crafts designs are fashionable today because they no longer depend on labor-intensive, expensive, hand-wrought techniques.

As I've said in the introductions to several other chapters in this book, innovations in construction technology—new sheet-goods, new adhesives, new fastening systems, new waterproofing materials—present carpenters with fresh possibilities. To be successful, a finish carpenter must develop an awareness of classical precedent yet must simultaneously maintain a keen understanding of contemporary materials and techniques.

ARCHED JAMBS

Historically, arched jambs have been built in millwork shops and many are still constructed using two specialized shop techniques. In one method, called *bricking*, a radius jamb is laminated from many small pieces of solid jamb-stock, each of which is cut with a band saw or a router and template (or often both) to the exact same radius. These small sections—like puzzle pieces—are all glued together, then milled and sanded *(Figure 13-2)*. The second method, *bending*, is accomplished by first either steaming a single piece of jamb or laminating several thinner pieces. Then the stock is

Figure 13-1. *Doorways (and mantelpieces) define a home's architectural style. The Georgian open-and-scrolled pediment above this doorway is mirrored by a similar pediment above the distant overmantel.*

Figure 13-2. In "bricking," one traditional technique for shaping curved doorways, small sections of the radius are first cut on a band saw, shaped with a router and template, and then glued together (top). Once the lamination dries, the assembly is milled and sanded (above).

bent around a purposefully constructed form and clamped until the curve sets. Because of the special equipment required, steaming is rarely done on job sites, and though bricking arched jambs can be accomplished on site with a jigsaw, router, and templates, that method is primarily used in a shop environment, too. Exterior jambs, which must be rabbeted and fastened together securely to accommodate weather and harsh use, are especially suited for a millwork shop, where the space, worktables, and equipment for such "specialty work" are available. That leaves laminating as a method that can be adapted as an on-site technique.

In the area where I work, a typical 3-ft.-wide and 8-ft.-tall arched jamb built by a millwork shop can cost anywhere from $600 to well over $1,000, depending on the width of the jamb and the wood species. To save money and to reduce pressure on imperiled species of hardwood, I build most interior arched jambs on the job site, whether they're made for doors or just open passageways, whether they're made from paint-grade or stain-grade material. I save money, particularly on jambs for extra-wide walls, because I don't laminate, plane, and sand wide boards of solid lumber. I prefer to use MDF or MDF-core plywood for extra-wide walls and arched jambs.

Figure 13-3. Installation usually begins at the top of the jamb — especially with an arch. But for a Palladian paint-grade arch like this, it's easier to set the side jambs and the flat head-jambs first.

Figure 13-4. *On paint-grade Palladian jambs, the arch can run past the bevel on the flat head. Cut the excess off with a circular saw, but leave it a bit long.*

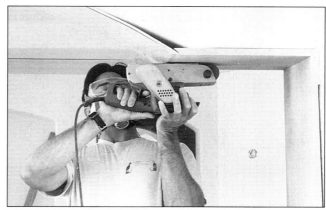

Figure 13-5. *Sand the joint until it's perfectly flush and smooth.*

Figure 13-6. *Brace the piece tightly into the center of the arch, and then countersink pilot holes for screws to secure the first layer to the drywall and backing.*

INSTALLING THE ARCH

I prefer having all arches wrapped with drywall before I start work. Even before the drywall is hung, I make a special point to check the framing and confirm that the framers installed sufficient backing throughout the span of the arch. I like solid backing, spaced no wider than 1 in. apart.

Trial and error has taught me that it isn't necessary to purchase expensive bending plywood, except for radiuses tighter than 2 ft. For most of the paint-grade arches I build, I use two pieces of ³/₈-in. MDF; for stain-grade arches, I use one piece of MDF for the first layer, and a piece of MDF-core hardwood plywood for the finished layer. I rip the material ¹/₈ in. wider than the widest part of the wall, which allows a little wiggle room just in case the sheet wanders during installation.

I usually start working on a jamb at the head, especially an arched opening. I find that it's easier to cut the legs at a miter and fit those miters to the arch rather than to cut the arch to fit perfectly between the legs. By installing the arch first, I'm able to concentrate on bending the pieces of material evenly and securing them tightly against the drywall and wood backing.

But rules are always broken. One exception is a paint-grade Palladian arch (Palladian arches are flanked by rectangular openings as in *Figure 13-3*). For the installation in this example, it was easier to install the jambs first, especially the flat head-jambs. Next, the arched head-jamb — which was cut long and ran well past the flat head-jambs — was secured to the drywall. Once we fastened the arch to the drywall, and glued and nailed it to the flat head-jambs, we cut off the excess material *(Figure 13-4)*, still leaving the arch a little long. Finally, we sanded the tail of the arch flush with the head-jamb *(Figure 13-5)*.

Measuring the circumference of an arch. Tape measures don't work well for measuring the circumference of an arch. Before I started using a Construction Master Pro® Calculator (see Chapter 11, page 160), I relied on a piece of flex-trim molding or a wide piece of tape. I still use duct tape frequently — it adheres well enough to the drywall without causing damage, and it doesn't stretch when it's removed. Simply stick the tape to the drywalled arch and mark the ends for length. Then transfer the tape to the first sheet of material and transfer the marks. Measure and mark the first sheet before installing it, which makes it easier to determine exactly how much longer the second piece

Figure 13-7. *If the drywall backing is inconsistent or not an even arch, shim the first layer so that it follows a smooth radius before installing the next layer.*

Figure 13-8. *Use a long level to position the second layer perfectly, so that both ends of the arch are level and parallel with adjacent jambs or arches.*

Figure 13-10. *Use a few brad nails to temporarily secure the joinery. Remember that glue and adhesive provide the permanent bond.*

Figure 13-9. *Use a short piece of stock, cut with a miter, to check that all the pieces are parallel and in a straight line. Otherwise, all the miters might not line up.*

needs to be — even to the short point of a miter.

A brace makes the job easier. As I've said before, the challenge of carpentry isn't in knowing the exact installation steps. More often the difficult part of our job is controlling awkward, heavy, or huge materials—like sheet goods and doors. Rather than struggling with a long wobbly sheet of plywood overhead, especially one that's coated liberally with adhesive, use a brace to secure the material.

Cut the brace 1/2 in. short of the opening, and then use a pry bar and shims to force the brace against the sheet and push it up tightly against the drywalled arch. Because the first layer isn't exposed, countersink pilot holes and use drywall screws to draw the sheet snugly against the drywall **(*Figure 13-6*)**, especially on tight radiuses. Nail

Figure 13-11.
To accommodate a post flanking the arch, create a two-piece head. Tack one piece around the post temporarily (right). Then join both pieces together before fastening the miters permanently (far right).

guns are often all that's needed on a wide arch. If the drywall backing is uneven, be sure to shim the first layer of the arch *(Figure 13-7)*.

Cut the last layer precisely. On a single, paint-grade arch, the length of the second and final layer is easy to determine if the length of the first layer is measured before it's installed. If the first layer comes out slightly short of the trimmers, simply add that amount to the measurement for the second layer. In truth, even the length of the second layer doesn't have to be exact — the jamb legs will cover at least 3/4 in. at each end of the arch.

But for stain-grade Palladian arches (like those in many of these photographs), determining the correct measurement the first time is essential because each side of the arch must be joined with a tight miter to a flanking head-jamb. I use a long level to help determine the exact amount that must be added to the measurement of the first layer. Then, after cutting the second layer, I brace it into the opening (a protective layer of cardboard prevents the brace from damaging the hardwood veneer) and use the level again to adjust the long points of the miters so that all head-jambs will be level *(Figure 13-8)*.

Next, I use a short piece of scrap cut with a miter to check that all the pieces are parallel and in a straight line *(Figure 13-9)*. If successive arches and head-jambs aren't in a perfectly straight line, the miters will be extremely difficult to cut and fit.

Once the piece is positioned perfectly, I use a brad nailer to fasten the glued miters. Brad nails are nearly invisible, but take the time to place each nail carefully and try to keep the number of nails to a minimum *(Figure 13-10)*. I leave the brace in the opening until the adhesive dries completely.

While developing this technique for laminating interior arched jambs, I learned through several mistaken attempts that adhesive-laminated arches bond permanently. I expect you can imagine what I mean without photographs of that mess!

Installing the Jamb Legs and Heads

Once the arch is installed, the exact dimensions of the flanking head-jambs are easy to determine and much easier to install than the arches. Miter the heads to match the miters on the arches, and then use a generous amount of glue and adhesive to ensure tight, long-lasting joints. On this job the two center heads had to be notched around structural posts. I tacked the first half of the head-jamb in place temporarily, and then used splines to join the two pieces (see "Slot Cutters and Spline Joints," page 203) before securing either miter permanently *(Figure 13-11)*.

Some carpenters have little respect for MDF, especially when it's used for a door jamb, but I often use MDF for thick-wall jambs, whether doors are being installed in them or not. I've found that if the jamb is shimmed solidly behind each hinge, and if the hinge screws are long enough to penetrate both the jamb and the trimmer, then MDF makes a durable and inexpensive thick-wall jamb.

I rip all the 3/4-in. jamb material to the correct width and ease the edges with a 3/16-in. roundover bit in my laminate trimmer (Figure 3-17, page 34). Like most jambs, the leg and head should be assembled before being installed in the opening to ensure a tight joint. I don't always rabbet the leg to

Figure 13-12. *Fasten the vertical jamb legs and horizontal heads together, and then work the miter first before nailing off the leg.*

secured, I plumb and shim the jamb leg, then fasten it to the trimmer with 2½-in. nails.

Finishing the Edges

The edges on stain-grade MDF jambs need special attention. On paint-grade jambs, I just sand the edges smooth, which is quick work because only ¼ in. of the jamb will be exposed after the casing is installed. But on stain-grade jambs, like the mahogany jamb in *Figure 13-13*, I use iron-on wood tape, the same species as the face of the jamb.

There are two things to remember when applying iron-on wood tape: Don't hold the iron in one position for too long, or the tape will burn. And when cutting pieces around an arch, keep the cuts radial to the curve of the arch, or the joints will look silly. Once the pieces are applied and all the miters fit perfectly, use a laminate trimmer to cut the tape flush with the face of the jamb. If you're terrified of a helper ruining the jamb with a laminate trimmer, 100# sandpaper works well, too, if a little slower. Once the edge is flush, use a block of soft wood to burnish the corner and gently ease the edge.

the head—drywall screws reliably secure a butt joint here *(Figure 13-12)*. After temporarily tacking the leg to the wall, I concentrate first on the miter between the flat head and the arched head, which must be glued and worked until it fits perfectly (as in Figure 13-10). Once the miter is

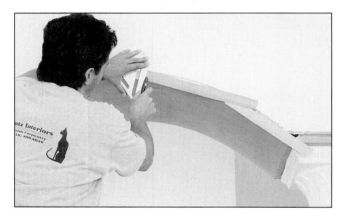

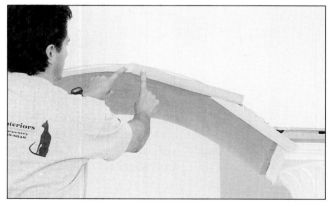

Figure 13-13. *After sanding the edge of the jamb with fine-grit paper and cleaning it of dust, apply iron-on tape (left). Lap the pieces of tape all the way around the arch, and then cut through the splices with a sharp utility knife (top right). Make sure the miter cuts are radial, that is, perpendicular to the face of the jamb. Remove the excess tape and press the two pieces together (bottom right). Then secure them with a little more heat from the iron.*

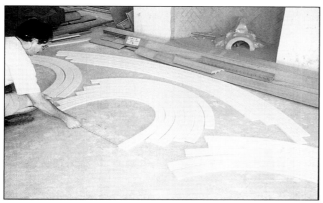

Figure 13-14. *To ensure that the correct piece is used at each opening, separate all flex-trim and label it before installation.*

ARCHED CASINGS

There are two types of casing that can be used for arched windows, doors, and doorways: solid wood casing must be used for most stain-grade work, but new flexible moldings can be used on paint-grade work. First I'll discuss easy installation techniques for the latter, more recently developed material; then I'll cover two systems for making stain-grade arched casing, along with a brief lesson in geometry (made easier by a Construction Master® calculator).

Paint-Grade Radius Flex-Trim Casing

Flexible polyurethane moldings (flex-trim) have helped to reduce significantly the price of labor and material for radius casing. I use flex-trim for almost all but the finest stain-grade installations. Urethane moldings are fairly simple to install, they take paint well, and they can also be ordered with an embossed wood grain for stain-grade or faux-finish applications.

Order material. To order radius molding, measure the run and the rise of the arch, especially if the opening is a segmented arch (more on segmented arches later). The molding manufacturer will do the math necessary to determine the exact radius so that the piece will match the opening precisely. Measurements for flexible molding don't have to be precise — urethane moldings allow for a lot of wiggle room, literally. But measuring for stain-grade solid radius molding is much different. I'll discuss that in a moment.

Order the job site. I rely on two simple techniques for fitting and installing radius trim. Each technique depends on the style of the door case-

Slot Cutters and Spline Joints

Biscuit joiners may be the rave, and thousands of cabinetmakers and finish carpenters swear by them, but I swear at them. Maybe I just haven't developed the right touch for a biscuit joiner. Often the two pieces I'm working with don't register exactly flush, and to me that's the most important task for the tool.

Then again, maybe I'm just a dinosaur because I still use my router with a slot cutter to make spline joints. (This isn't the first or the last time I'll demonstrate my dependence on spline joinery). Tightly-cut spline joints are unrivaled in their ability to instantly register two pieces of wood exactly flush and to create a strong and durable joint, especially when laminating two boards into one wide board *(see photo)*.

I set the depth of my router so that the slot cutter is slightly off center. That way it's clear which direction the two pieces must be joined. I always hold my router on the face of the material, and then I mark a large X on the back to further ensure that the pieces are joined and installed in the right direction. I prefer to use a 1/4-in. slot cutter for spline joints because 1/4-in. MDF fits into the slots snugly and guarantees flush joints.

I prefer spline joints to biscuits. When laminating two veneered boards into a wide board, the faces must be perfectly flush.

work — whether rosette blocks are being installed or the corners are being mitered. Regardless of the installation technique, I always begin by laying all the material on the floor separated into like piles, especially if the job has several arched openings *(Figure 13-14)*. That practice prevents carpenters from grabbing the wrong piece for the wrong opening.

Installation with rosette blocks. Mixing arched casing with rosette blocks can pose a design dilemma. Should the rosettes be placed as they are on a rectangular opening — with the bottom of the

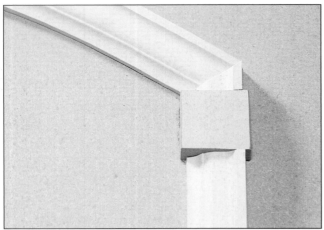

Figure 13-15. *If a rosette is installed beneath the head of the jamb, then the radius casing must be returned to the top of the rosette with a miter.*

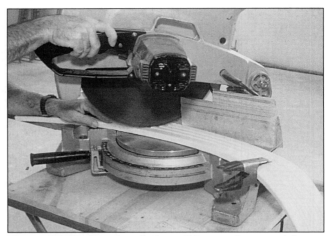

Figure 13-18. *Flex-trim wiggles as if it were alive, especially when it gets near a saw. While cutting, use a clamp to hold it still.*

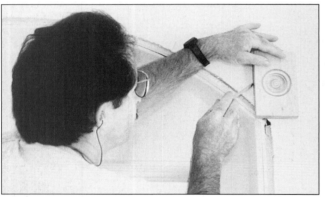

Figure 13-16. *No matter what the casing design—whether the rosettes are placed above the head-jamb or below—always begin by tracing the location of the rosette.*

Figure 13-19. *Once the casing is cut to size and fastened to the wall, hold the rosette in position and mark the amount that must be trimmed from the top and bottom—smaller radiuses require taller rosettes. The rosette should be at least 1/8 in. taller than the casing.*

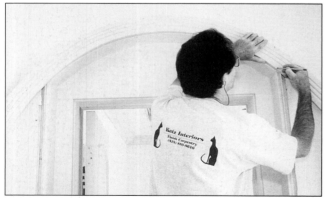

Figure 13-17. *Temporarily fasten the casing to the jamb. Use as few brads as possible, but enough to ensure that the casing follows the radius of the jamb evenly. Then mark each end by transferring the line previously traced along the mating edge of the rosette.*

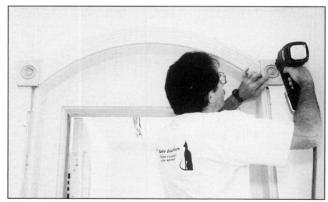

Figure 13-20. *Fasten the rosette to the jamb and the wall using nails and adhesive caulking.*

Figure 13-21. *Carefully mark the points where the moldings intersect. To ensure a tight-fitting joint on the first attempt, use a square or straightedge and be certain your marks are in a perfectly straight line.*

Figure 13-22. *If there's no room for a straightedge, mark both pieces at the top and bottom of the intersection, and then take them down and extend the marks into straight lines using a square.*

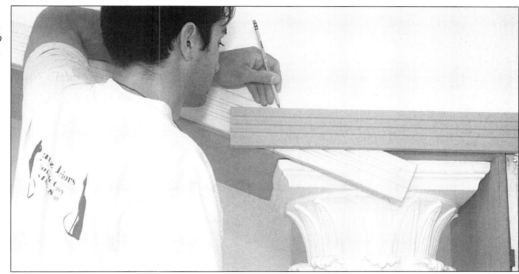

Figure 13-23. *Make a template for irregular radius jambs, rather than depending on a mathematically determined arch. Use a thin piece of plywood and trace the radius of the jamb, and then note on the template the amount of reveal for the radius.*

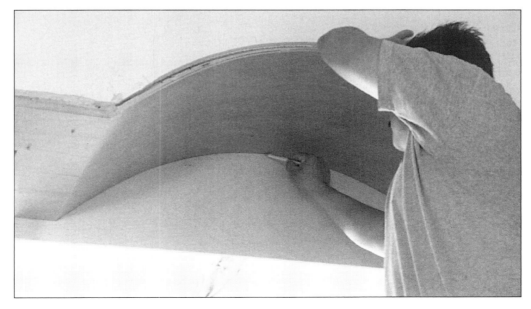

DECORATIVE DOORWAYS

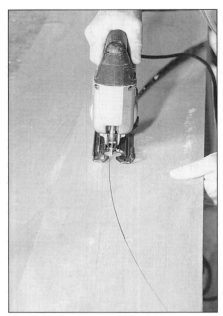

Figure 13-24. *Carefully cut the template outside the line of the reveal, and then use a belt sander to finish the cut right up to the line.*

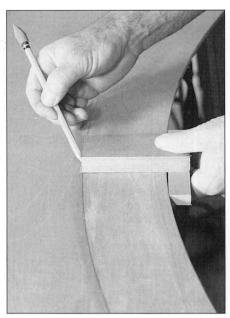

Figure 13-25. *A quickly built jig and fence makes it easier to trace the outside edge of the radius molding.*

Figure 13-26. *Use a jigsaw and belt sander to make the second cut, too, then temporarily attach the template to the molding stock.*

rosette nearly flush with the head of the jamb? Or should the rosettes be dropped down and installed with the top of the block flush with the top of the jamb *(Figure 13-15)*? In either case, a little extra "trim" carpentry is always required.

In a home we recently completed, the designer wanted the rosette blocks on the arched openings positioned exactly as they were on the rectangular openings — with the bottom of the block almost flush with the bottom of the head-jamb. Unfortunately, mitering the radius casing to a stock rosette couldn't be done because the length of the miter was longer than the height of the stock rosettes.

Solving that problem was easy — I ordered custom rosette blocks (see "Installing Rosettes and Plinth Blocks"). I laid out and installed the radius casing, after which I cut the blocks down — an equal amount from the top and the bottom — to fit each arched opening *(Figures 13-16 to 13-20)*.

Installation with miters. Radius casing can also be mitered at the corners, or mitered at flanking rectangular openings (see "Radius Casing with Keystones," page 208). I use the same technique for marking and cutting these miters as I do for decorative ceiling trim (see Chapter 12, "Flex-Trim Corners," page 187).

Start by fastening all the molding to the jambs. Leave the ends long so that they overlap each other, and use only enough brads to secure the material temporarily to the jamb and wall *(Figures 13-21 and 13-22)*.

Stain-Grade Radius Casing

Flex-trim can be used for stain-grade installations, too, but it's difficult at best to match stain colors between urethane and hardwood moldings, unless both materials are being faux finished. For true stain-grade installations, I opt for solid-wood radius casing.

I use two methods for making these costly pieces of molding. One method is for irregular radiuses and the other is for perfect radiuses.

Irregular radiuses. Some radius jambs have irregular shapes, even though they're supposed to be — and might even seem to be — smooth arcs. Irregular radiuses aren't surprising given the framing on some homes, but also consider that the drywall follows the framing, and the jamb (built-up from several pieces of 1/4-in. or 3/8-in. MDF) follows the drywall. Elliptical radiuses often fall into this category. In such cases, I make a template for each piece of trim *(Figure 13-23)*, and then use the template to cut the molding.

I start by cutting just outside the line of the tracing with a jigsaw *(Figure 13-24)*, and then use my belt sander to remove the remaining material right up to the line. Once the arc is smooth, I cobble together a quick jig to trace the second arc at the top of the molding — in this case 3 1/2 in. from the first arc *(Figure 13-25)*. A set of scribes will accomplish the same task, but the line is difficult to scribe without wiggling. The short fence on the jig ensures that

Installing Rosettes and Plinth Blocks

When I case a doorway with standard molding, I always install the head casing first because it's easy to center the head. The placement of the miters on the head casing determines the location of each piece of leg casing. But for rosettes and plinth blocks, I take a different approach.

On rectangular jambs, I usually start with the rosettes and install them on the jamb and wall with adhesive caulk and nails — latex caulk works well because it adheres aggressively to the drywall, and it's also easy to clean up.

Since the jamb in this example had a miter on the left side, I traced the location of the rosettes — rather than just nailing it in place *(A)*. I marked and fit the head casing next *(B)*. Then I installed the rosette *(C)* and, finally, the head casing. The head casing can be installed first, and then the rosette pressed tightly against the casing, but whichever method you chose, always glue the joint between the rosette and the casing.

After the rosettes are installed, I turn to the plinth blocks. The same combination of latex adhesive and nails also secures the plinth blocks *(D)*. While the adhesive is still wet, I make sure to cut and install the casing leg *(E)*. I apply glue to the top and bottom of the casing, tack the casing to the jamb, and then use my pry bar to lift the plinth slightly and squeeze all the joints tightly closed.

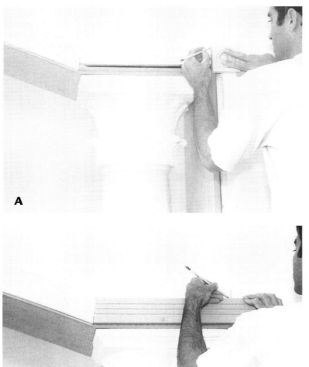

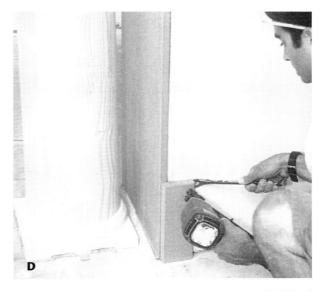

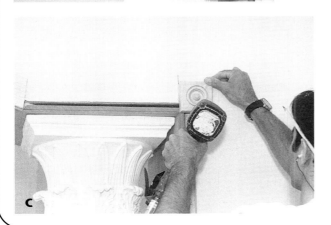

Decorative Doorways

Radius Casing With Keystones

Another common application for radius casing, also found in many architectural styles, including Gothic, Georgian, and Provincial, are keystones. I often make my own keystones on the job site, at least if there's no carving or other detail in the rosettes *(Figure A)*.

The flexible casing around this doorway was almost 3/4 in. thick. I cut blocks from 1-in. MDF and cut 22 1/2-degree angles on both ends. Because the blocks were more than 1/4 in. thicker than the casing, I was able to chamfer the edges all the way around, using a router and chamfer bit *(Figure B)*. To increase the depth and shadow of the keystones, I cut a second block from 1/2-in. MDF, which I laminated on top of the 1-in. block. The second block was cut 1/4 in. small, all the way around, and chamfered almost the full thickness of the piece *(Figure C)*.

Figure A. *Keystones are another popular trim detail with radius casing, and they can be made easily on the job site.*

Figure B. *These keystones are made from two pieces of MDF. The 1-in. MDF base is cut with two 22 1/2-degree ends, and then all four sides are chamfered.*

Figure C. *The 1/2-in. MDF top is cut in exactly the same shape but 1/4 in. smaller than the base. The top is also chamfered, almost full thickness, on all four sides.*

the second arc will follow the first without wiggling.

I repeat the same cutting/sanding technique for the second line *(Figure 13-26)*, and then attach the finished template to a piece of rough stock. The top-bearing pattern bit in my router does the remainder of the work *(Figure 13-27)*. Cutting flutes is easy work, too, with a router and micro-adjustable edge-guide *(Figure 13-28)*.

Perfect radiuses. I use a different, far more productive technique if the jambs follow a perfect radius. This is almost always the case with trim for manufactured doors and windows. First I find the radius of the arched jamb. For circle-top jambs, or jambs which are the width of a circle's diameter, the radius is simply half the width of the jamb opening. But for segmented arches—an arched opening that is narrower than the diameter of the circle, a little math or a calculator is necessary.

Figure 13-27. Use a router and bearing-guided pattern bit to trim the stock to size.

Figure 13-28. The ultra-fine adjustability of a Micro Fence® attachment simplifies the task of laying out and cutting the flutes.

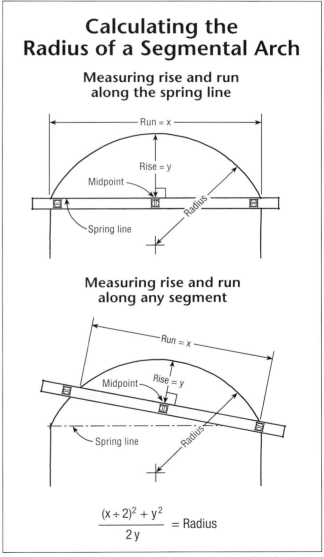

Figure 13-29. To find the radius of a segmental arch (part of a circle), use the formula above. You can measure the rise and run along the spring line (at top) or along any line cutting through the arc. The results of the formula will be the same.

I use the same method that I described a few pages back for ordering flex-trim molding. The only difference is that for stain-grade material, I take the measurements very carefully. First I determine the rise and run, or bottom cord, of the arch. The rise and run must be measured precisely. Even a difference of 1/8 in. will have an impact on the fit of the finished piece.

The formula for calculating the radius of a segmented arch is easy to follow *(Figure 13-29)*, but feet and inches must first be converted into decimal equivalents. A Construction Master Pro® calculator will do that job, and it will also do the geometric formula, which saves me a lot of time. Besides, this is one job where even a small error can be disastrous, so a calculator becomes a life-saver.

On a Construction Master Pro®, simply enter the dimensions of the rise and run—in feet and inches or just in inches. Then press the convert key followed by the Diag/Radius key and the radius will appear on the screen.

$$\frac{(x \div 2)^2 + y^2}{2y} = \text{Radius}$$

Micro Fence

A circle-cutting attachment is also available for the Micro Fence® router accessory, and so is an accessory for cutting an ellipse. These tools are not cheap—in other words, expect to pay exactly what a finely engineered instrument should cost *(Figures A, B)*.

Figure A. *The circle-cutting trammel arms are available in different lengths, which allows for cutting some truly massive radiuses.*

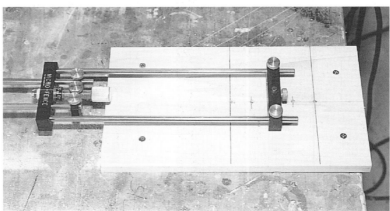

Figure B. *For the repetitive work of cutting radius casing with flutes, I lay out each radius point on a trammel block. Then I adjust the length of the entire trammel arm to match the required radius.*

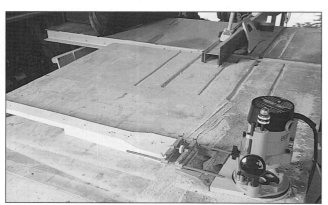

Figure 13-30. *Including a sub-base at the end of a trammel arm is one way to simplify attaching a plunge router. The trammel arm in this example is adjustable.*

To determine the length of the molding required to cover the arc, press the Circ/Arc key three times. The diameter of the arch will appear the first time, the area of the arch will appear the second time, and finally, the circumference of the arch will appear the third time. I use that figure to determine the rough length of the board I need to use (for safety's sake, I add an inch to that dimension), but I use the calculated radius to set up my cutting jig.

Extension arms for swinging arcs. It wasn't long ago that I learned how to mill radius molding with a plunge router and circle-cutting attachment. I was immediately surprised by the simplicity of the setup and the ease of the job. The circle-cutting system is simply a long arm attached to a plunge router. There are several methods for attaching the router to the arm. Friend and fellow carpenter Steve Phipps taught me how to mount a router to an arm with a large flat base. Steve fastens the router onto the wood sub-base, and then goes to work *(Figure 13-30)*. Micro Fence® (see Appendix 1) manufactures a circle-cutting attachment that's not only a pleasure to use but also a beauty to behold (see "Micro Fence," above). I've also combined a Porter Cable accessory fence with an extended arm for my PC 7529 plunge router.

No matter which method or arm you choose, it is essential to have at least a little adjustability in the

Figure 13-31. A stock accessory fence makes a good starting point for an extension arm. Any length of additional wood can be used, but an accessory fence allows much-needed adjustability for exact fine-tuning.

Figure 13-32. Cut the extension arm long enough to allow for the full range of necessary trammel points. Remember, the initial trammel point is the shortest radius (for the inside of the curve); the second trammel point is always farther away. Notice that the width of the bit must be added to the length of the radius because it's the far edge of the bit that cuts the inside of the radius.

Figure 13-33. Use the same trammel arm for cutting flutes, just add additional trammel-point holes in the trammel block. Use a test piece to fine-tune the layout before attempting to cut the molding.

length of the arm, so that the initial radius—the radius at the reveal line—can be dialed in accurately. Once the initial radius is set from the starting trammel point, additional trammel-point locations in the tail of the extension arm determine the position for succeeding cuts — the overall width or height of the molding, the location of flutes, etc.

Swinging the arc. Though it may appear to be difficult, cutting radius molding with an extension arm is easy repetitive work, if the work surface is prepared properly. Any sturdy, large-surface worktable will serve, so this task is definitely not reserved for a millwork shop.

Before attempting to cut an arch with a router, be sure to secure the work piece in a form or jig. In this example, the jig is made $11^{1}/_{4}$ in. wide for standard 1x12 material. But fasten all the pieces with screws, so that it's easy to alter the size of the jig for different radiuses and different widths of material.

A manufactured accessory fence complements the site-made extension arm **(Figure 13-31)**. Any piece of material will work well for the actual arm, though the stiffness of a hardwood 1x4 is helpful; hardwood also provides better purchase for fastening the accessory fence to the extension arm. Just be sure to make the arm long enough to allow for the initial trammel point location as well as the farthest point necessary to cut the full width of the molding. Locate the approximate position of the initial trammel point, but remember to measure from the far-side of the router bit—that edge of the cutter is the inside edge of the radius molding. Next, drill the second trammel-point location. To

DECORATIVE DOORWAYS 211

Figure 13-34. *To install casing over a Palladian arch, temporarily fasten the casing to the jambs, and then trace the overlapping joints. Next, cut all the miters, but leave the miter on the flat head a little long. Install the radius heads first, and then install the flat head. If the joints were traced carefully, it should be necessary to shave only a sliver from the miter for a perfect fit.*

Figure 13-35. *Casing can be decorated by applying ogee molding to the face (far left). Held back from the edges and the ends of the casing, this ogee detail picks up the paneled look of the library (left).*

locate that point, add the width of the molding to the width of the router bit—notice that the inside of the bit cuts the wider, outside radius *(Figure 13-32)*.

For cutting flutes, locate intervening trammel-point holes between the inside and outside radius points at the butt-end of the trammel arm. But before cutting any material, go through a non-powered practice run to check the location of each trammel point. Cutting a piece or two from a scrap of MDF is also a good way to fine-tune the trammel points for cutting flutes. I always make several pieces of radius trim from scrap material, as I fine-tune the layout *(Figure 13-33)*.

Installing radius casing. Whether the trim is paint-grade or stain-grade, the same installation techniques apply, except that for the latter you work a lot more slowly and carefully. On a single radius head, start by temporarily fastening the uncut casing to the jambs, both head and legs, overlapping where you'll cut the joints. Trace the joint lines *(Figure 13-34)*, cut both the miters, and cut the miters on the legs, too, but leave them a sliver long. Install the head first, then hold the miters on the legs against the miters on the heads, and shave the miters on the legs until the joints match perfectly.

BUILT-UP CASING

Radius head-jambs aren't a prerequisite for dramatic doorways. Neither is custom one-of-a-kind molding

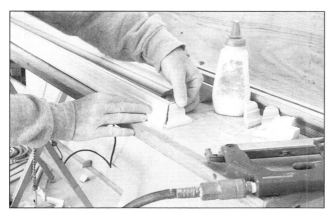

Figure 13-36. *Preassemble everything, especially the end returns. Use glue and brads on every joint.*

Figure 13-37. *Support the top shelf at the wall with a back-out on the inside of the crown molding. For large crown molding, install the back-out support (or ledger) on the wall, and then slip on the preassembled over-door molding.*

always the best way to enhance a doorway. Simple casing, based on fine classical details, is often the best choice for dressing up any doorway.

Colonial and Victorian casings were often embellished by a back board molding, but with a router and simple ogee bits, a carpenter can add interest to any flat stock.

To highlight the paneling detail in a library job, we used 1x4 mahogany for the casing, and then applied a plant-on ogee molding *(Figure 13-35)*. But an ogee profile is only one of many types that are easy to cut with a router. Beading bits are available in a large assortment and can add detail to the edge of any casing.

OVER-DOOR TRIM

Georgian doorways—which inspired designs for many Colonial homes — often included over-door trim. Raking crown molding, known as pediments, often topped elaborate entablatures above 18th-century doorways. The molding in some pediments followed strict classical lines and rose above a tympanum (triangular gable end above a Greek entablature) to meet in the center above the doorway. Even more elaborate pediments were open and separated by an additional decorative element—an urn, a lyre, or a large shell (Figure 13-1). The upper ends of the crown molding were often embellished with scrolled or swan-necked crown molding.

Miniature Mantelpieces

But over-door trim doesn't have to be elaborate or expensive. Instead, simple moldings can be combined to form an inexpensive yet elegant design that's easy to install. Over-door trim is often assembled just like a miniature mantelpiece, from three pieces of material: soffit, cornice, and top shelf.

First the soffit. Start by ripping and crosscutting the material for the soffit. Measure the outside length of the casing head or frieze and measure the thickness of the casing, too, and then add enough for an overhang. In other words, the soffit should be slightly proud of the casing, both in the front and on the sides. An even 1/4-in. overhang is sufficient all the way around, though 1/2 in. also works well. Anything broader than 1/2 in. can add too much drama to the doorway, but appearances are subjective, so have your client make the final decision.

Then the crown. Depending on the type of molding, the soffit might also be formed by the base of the crown molding. The solid-back crown molding I chose to use in the home pictured in the next few photos works especially well for over-door trim because a separate soffit isn't required. All that's needed is backing.

I added 1/2 in. to the overall measurement of the head casing, for 1/4-in. overhangs on each end. There were several doors of the same size in the entryway, so I used a stop on my miter saw to speed up production. The casing measured 3/4 in. thick, so I cut the return pieces 1 in. long, to allow for the same 1/4-in. overhang on the front. Remember, those are all short-point measurements for outside corners, except for the butt cuts on the returns.

I applied the ends using miter clamps, glue, and brads *(Figure 13-36)*, then backed-out the center with a piece of MDF to add fastening support for the top shelf *(Figure 13-37)*.

The top shelf adds drama. On a mantelpiece, I often add an additional piece of nosing to the

Figure 13-38. *After beading the top shelf, place the crown molding beneath it, and then glue and fasten the two pieces with brad nails.*

Figure 13-39. *Fastening the completed molding to the wall is quick work. Shoot 2¹/₂-in. nails through the bottom of the crown and toenail the top shelf to the header.*

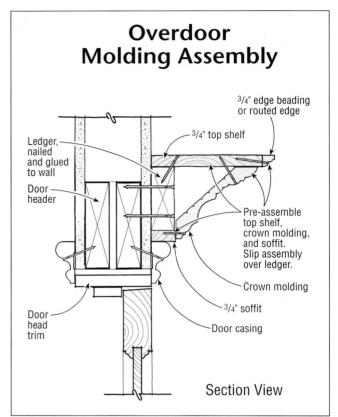

Figure 13-40. *Over-door trim can be preassembled just like a mini-mantelpiece, then installed on the wall above each doorway.*

edge of the top shelf, but on these miniatures I wanted to keep things simple. Instead, I milled the top shelf ¹/₂ in. longer and ¹/₄ in. wider than the projection of the crown molding, and then ran a beading bit on the front and side edges. To add a little harmony, I used the same beading bit that was used on the edges of the casing *(Figure 13-38)*.

Installation. The over-door profile in this home was so small that 2¹/₂-in. finish nails were more than long enough to penetrate the bottom of the crown and reach the header *(Figure 13-39)*. I also toenailed the top of the shelf to the header. But installing this molding often requires another step.

For large over-door profiles, I install a ledger on the wall, rather than backing out the inside of the crown molding. I rip the mounting ledger from a piece of scrap so that the ledger fits snugly but not tightly inside the crown molding. I crosscut the ledger about ¹/₈ in. shorter than the length of the soffit, to avoid any problem with the short points of the crown clearing the ledger.

Large over-door designs almost always include a soffit made from ³/₄-in. material, so I use a piece of ³/₄-in. material to lift the ledger above the head casing or frieze, which allows space for the soffit to slide in. Once the ledger is secured to the wall with a liberal amount of adhesive caulk and nails *(Figure 13-40)*, I slide the three-piece over-door assembly on top of the ledger, slipping the soffit between the casing and bottom of the ledger. Then I nail right through the top shelf.

Figure 13-41. *Manufacturers often ship columns ready to install, but these wooden shafts needed to be trimmed flush at the astragal before installation (far left). A circular saw made the cut, and a belt sander finished the job (left).*

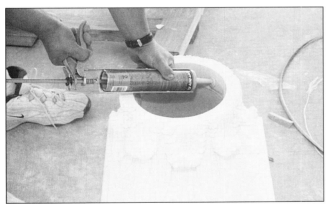

Figure 13-42. *Urethane glue is the best adhesive for urethane foam products, such as these capitals. Use a generous amount of adhesive, but try not to apply it where it might squeeze out and create a clean-up problem.*

COLUMNS AND PILASTERS

The most dramatic decorative element in a classical doorway is a pair of columns. Columns come in many different sizes, shapes, and styles. When they're rectangular, it's proper to call them pilasters. An "engaged" column or pilaster is one that's installed against a wall or doorjamb and seems partly hidden in the wall.

Because they're rectangular, pilasters are easy to build from stiles, rails, and flat, recessed panels (see Chapter 11, "Decorative Walls"). But columns aren't so easy. Admittedly, I have an addictive thirst for fine tools, but even I don't carry a 20-ft. lathe in my van.

Fortunately, there are several manufacturers who market columns today, and though they're expensive, manufactured columns are not beyond the budget of most custom homes. Like modern moldings, columns are now constructed from a variety of materials, including urethane foam, fiberglass, lightweight concrete, plaster, and even laminated wood for stain-grade applications. But no matter which material they're made from, they are all installed in much the same manner; the only difference is whether they're solid columns or split columns.

Solid Columns

Solid columns are just that; they're made in one piece, not designed to wrap around a structural support. Because there are fewer pieces, solid columns are much easier to install.

Trim the top. Columns manufactured from synthetic materials are cast in molds and they're shipped ready for immediate installation. Wooden columns sometimes require a little special attention.

For instance, on the job shown in ***Figures 13-41*** to ***13-44***, we trimmed the tops of the wooden columns right to the astragal — per the manufacturer's instructions — and then inserted a sleeve inside the column. We slipped the capital over the sleeve and secured all three pieces with adhesive caulk, screws, and nails.

Scribe for the base. Cutting the column at exactly the right height can be tricky. Even laying out a pencil line around the circumference of the

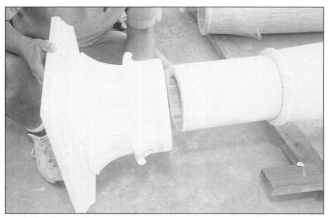

Figure 13-43. *The manufacturer supplied wooden sleeves that slipped inside the shaft. The capital then slipped over the sleeve.*

Figure 13-44. *To fasten the three pieces together, use screws wherever it's easy to fill the holes, but rely on nails near the fine details on the capital.*

Figure 13-45. *Place the column in position, then shim it off the plinth until the capital is snug against the jamb. Cut a scribe block exactly the height of the base, and then place a sharp pencil on top of the block and draw a line all the way around the foot of the shaft.*

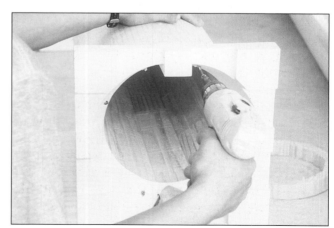

Figure 13-46. *Install the base on the foot of the shaft before installing the column. Screws are easier to hide in the bottom of the base.*

column can be a challenge. But by placing each column in position—without the base installed, then scribing the column to the plinth—and allowing just enough room for the base, we were able to solve both tricky challenges at once *(Figure 13-45)*. After cutting the column at the scribe line, we fastened the base to the foot of the column with adhesive and screws *(Figure 13-46)*.

Split Columns

Split columns are exactly what they're called—split in half so the two halves can be wrapped around a structural post. And it's not just the column that's split but the capital and base, too. Cutting the column to the perfect height is only part of the challenge. Often the real difficulty lies in assembling the two halves of the capital and the base so that the joint isn't visible.

Figure 13-47. *These posts had to be chamfered slightly to allow for the taper of the column.*

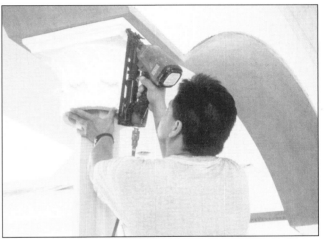

Figure 13-48. *Assemble the two halves of the capital and apply glue to the top of the capital before raising it to the head of the jamb (top). Tack the capital to the jamb, making sure that the jamb reveal is even around the perimeter of the capital, but don't fasten the capital permanently until the shaft is installed (above).*

Layout. Classical columns are patterned after two common figures used for vertical support — the human form (caryatids) and the form of a tree. The most commonly used forms taper, beginning about one third of the way up from the base. (A well-designed classical column also exhibits entasis, a slight convex curve along the length of the taper, which serves to offset the optical illusion that straight columns narrow at the center.)

Because the columns in *Figure 13-47* tapered and the structural post around which they were to be wrapped did not, the first chore in this installation was to chamfer the edges of each post from about two-thirds of the way up to the ceiling.

Install the capital. Once the inner back of the capital cleared the post, we were able to hold the capital against the head of the jamb and trace its final position. Then we assembled the two halves of the capital using polyurethane adhesive caulk. (The capitals were made from urethane foam. Urethane glue is the only reliable adhesive for urethane foam.) To draw the two halves together, in lieu of belt-clamps, we wrapped electrical wire around the capital, and then drew it snug by twisting a nail through a loop in the wire. We were careful not to twist the nail so much that the wire would cut into the capital — only enough to begin to press the adhesive evenly out of the joint.

While the adhesive set, we assembled the two halves of the capital around the post, applied adhesive to the top flange, and tacked it in place *(Figure 13-48)*. We postponed fastening it permanently —just in case the capital had to be shifted slightly to align with the column.

Scribe for the base. Fitting the column was the next step. As with the solid columns, we set each half in position and then scribed the bottom at the

DECORATIVE DOORWAYS 217

Figure 13-49. *Place each half of the shaft in position, then scribe the foot for the thickness of the base.*

height of the base. Blocks cut at exactly the height of the base ensured that the scribe line would be perfect the first time *(Figure 13-49)*.

After cutting both halves to length, we applied a heavy bead of urethane glue to each side, positioned the pair around the post, and used wire to clamp them together. Then we angled finish nails through the two long joints.

Next we applied urethane caulk to the top of the astragal, lifted the shaft, and slid the two halves of the base under the foot of the column. Minor adjustments to the base and the shaft ensured that they fit evenly on the plinth and the capital, respectively. Then we fastened the three pieces together permanently with finish nails.

Chapter 14

Bookshelves

A complete book could be written on bookshelves—there are so many designs, styles, and construction techniques. In this chapter I'll cover only three basic types of bookshelves, but by exercising a little creativity, the methods and styling used for each of these can be applied toward almost any bookcase design.

A CLASSICAL BOOKCASE

Like the doorway designs covered in the previous chapter, most bookshelf designs depend on the classical orders (see Chapter 1). Though modern, streamline casework might appear to have originated wholly in the 20th century, almost every bookshelf depends on a base, vertical dividers, and a cornice. These three components are analogous to the classical plinth, columns, and entablature. Understanding that basic principle allows carpenters the freedom to design and ornament bookshelves in any period style. The mahogany bookcase in *Figure 14-1* is built in the Arts and Crafts style, with hints of Gothic and Art Deco designs typical of that period.

Drawings

As with all aspects of our trade, success (productivity and profit) depends on careful layout. Because finish carpentry tasks are all repetitive, always invest time and effort in drawings. That way, milling and assembly can proceed without the distraction of costly errors. There are several construction characteristics that should be considered in every bookshelf drawing (*Figure 14-2*).

To rabbet or not. Decide whether or not to rabbet the back edges of the vertical sides for 1/4-in. plywood backs. Though the 1/4-in. rabbet isn't difficult to cut with a bearing-guided router bit, I often avoid that joint by adding finished sides to all end cases. I find that finished sides on the end cases are a necessity in order to balance the overhang of the vertical stiles (face frame). Only with an added finished side can the stile on the end be flush with the outside of the bookcase and still overhang the interior side of the case almost the

Figure 14-1. *Bookshelves, like doorways and mantelpieces, lend themselves to classical decoration. Details on stiles, shelf nosing, and cornice can be combined to mimic any period—like this Arts-and-Crafts example.*

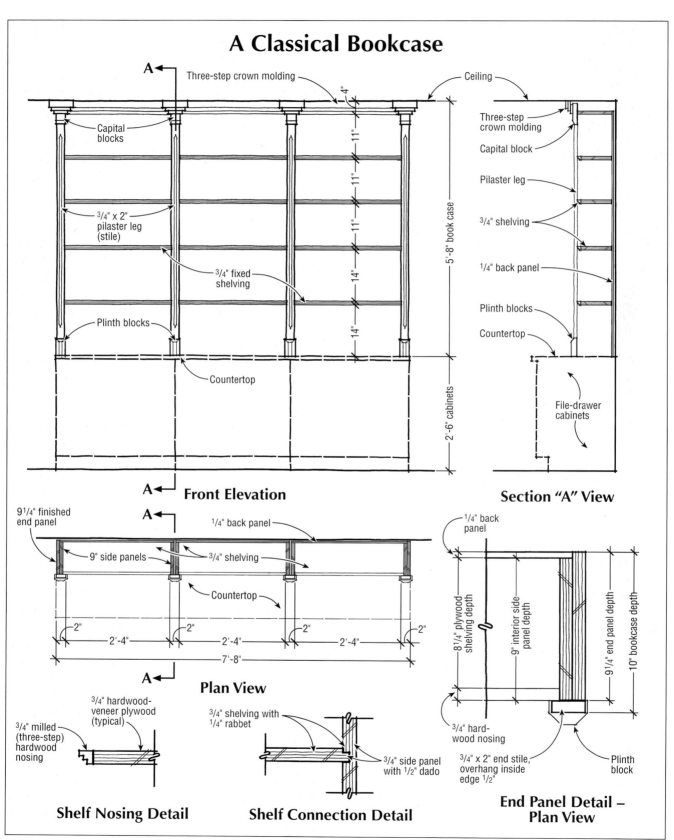

Figure 14-2. Rather than rabbet the vertical side panels to receive the 1/4-inch plywood backing, use a finished panel on each end, as shown in the End Panel Detail. Adding the finished end panels also makes the overhang of the end stiles more consistent with the inner "meeting" stiles. The fixed shelves are rabbeted for a precise fit into 1/2-in. dados. In general, 3/4 in. shelving for books should not span more than 30 or 32 inches.

Figure 14-3. *Install a separate toe kick for long runs of cases, so that it's easier to shim and level all sides of the toe kick, as well as find studs with fasteners. If the toe kick is installed perfectly level, the cases almost fly into place.*

same amount as the center "meeting" stiles (where two cases butt against each other).

To dado or not. Decide if the shelves will be dadoed into the sides (stationary); if the sides will be drilled for adjustable pegs (see "Holes for Adjustable Shelving," next page); or if the sides will be dadoed vertically for adjustable standards (see "Grooves for Adjustable Standards," page 223).

If the sides are dadoed for stationary shelves, decide whether the width of the dadoes will equal the thickness of the shelving, or if the shelving will be rabbeted to fit into a narrower dado. I prefer the latter because plywood thickness and router bits are rarely a match, but more on that in a moment.

Also, decide on the exact depth of the dado so that all dimensions (including shelving lengths) can be determined from the drawings.

Add a kick. If the unit sits on the floor (carpet or hard surface), decide on the height of the toe kick. Anywhere from 2 in. to 4 in. is standard. I prefer to make toe kicks separate from the case boxes, especially for long units composed of several cases. Leveling a separate toe kick is a snap: it's easy to reach all around a 3-in.-tall toe-kick frame and place shims and screws right where they're needed (*Figure 14-3*).

Allow for scribing. Bookcases that abut walls should be scribed to fit, so that additional molding isn't necessary to hide an out-of-plumb or bowed wall. Scribing a stile to fit tightly against a wall isn't difficult, but cases aren't often constructed with enough room between ends and walls to allow for scribing. I always make cases 1 in. narrower than the finished opening, and I allow the end stiles to overhang the sides by an additional 1/2 in., which provides plenty of room for scribing and also ensures that the stile widths across the unit will be nearly equal and symmetrical.

Decide on case and shelf depth. I build bookcases mostly from plywood (hardwood veneer with an MDF core), so some type of nosing is always required. I prefer to have the nosing tuck behind the face frame (as it should for adjustable shelves), rather than have the nosing terminate against the face frame. So I deduct the thickness of the nosing before ripping my shelving.

While on the subject of shelving depth, remember that there's no rule that bookshelves must be 12 in. deep. In many homes space is at a premium, so for cases that are meant solely for books, I make the finished shelves between 7 in. and 9 in. deep—any more and the shelves might become storage space for nick-nacks. Of course, if the shelves are meant for collectibles or other display, the depth of the shelf must meet the need.

Decide on height and width. If at all possible, bookcases—like all cases and cabinets—should be built-in and extend to the ceiling (as in Figure 14-1). If the top of the case stops short of the ceiling, it becomes nothing more than a dust shelf. But more importantly, from a design perspective, the cornice on the bookshelf should match the height of the room's cornice or ceiling. Otherwise, the two common forms of entablature compete inconclusively.

There are occasions where this rule is meant to be broken. If the ceiling slopes, if the ceiling is unusually high, or if the bookshelves are designed as moveable furniture, a second cornice is appropriate (as in Figure 14-20 on page 228).

Whether or not the case extends to the ceiling, if crown molding is used on the cornice, be sure to include the crown molding in the overall height of the case. Also, design the top rail so that it is flush with the face of the stiles and the crown molding can wrap continuously around the face of the bookcase (that rule can be broken, too, but more on that in a moment).

Furthermore, be sure that the top rail extends high enough to provide adequate backing for the

Holes for Adjustable Shelving

The easiest way to make adjustable shelves is to drill a series of holes in the case sides, and then use shelving pegs to support the shelves. Of course, one center shelf should be stationary and permanent, to prevent the sides from bowing and dropping all the shelves.

For years I've drilled holes for adjustable shelving using a home-made template cut from pegboard. Pegboard is full of perfectly symmetrical holes, and it's easy to rip a piece the same width as the case sides, and then use those holes as guides for drilling adjustable holes. A plain piece of hardboard also works well, especially for custom hole patterns *(Figure A)*. I run two wide strips of blue masking tape before laying out the holes, so that the layout is easy to see and easy to correct—it usually requires a few attempts before the layout is perfect.

But these days, when it comes to drilling holes for adjustable shelving, I prefer to use a Festool® plunge router (see Appendix 1), along with their cutting guide system, meant specifically for drilling 5-mm holes for adjustable shelving. The plunge router is ingeniously designed for smooth single-handed operation (the posts are offset from the handle). The accessory plate has a spring-loaded rocker switch that—also operated with a single hand—allows for quick release and repositioning *(Figure B)*.

Both the hardboard and the Festool® guide must be clamped to the work surface, though the Festool® guide is equipped with two strips of what I call "magic rubber." Though they're not tacky, these strips prevent the guide from moving, except in the most extreme cases. Still, clamps are good insurance.

Figure A. *Pegboard, which is full of perfectly symmetrical holes, makes a good template for drilling holes for adjustable shelving. For custom shelf layouts, I used to drill my own holes in a hardboard template.*

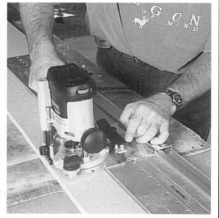

Figure B. *Now I use a Festool® guide and plunge router. This perfectly engineered system relieves the frustration of drilling so many perfectly spaced holes in verticals.*

crown, and low enough to provide both backing and reveal below the crown molding.

The width of bookcases, on the other hand, is often governed less by design than by physics. Shelving for books that is 3/4 in. thick can't span more than 30 in. or 32 in., or the shelves will sag. Longer spans call for thicker shelving or wider nosing or both.

Construction

Once I have a good drawing, I like to jump into building bookcases. Ripping the sheet goods comes first. After all the shelves and verticals are ripped to the proper width, I crosscut the verticals to the correct height and the shelving to the correct length. Then I lay out the vertical dividers.

Eliminate measurements with a story pole. Because the shelves on these cases will be stationary and dadoed into the sides, I use a story pole for positioning the dadoes, which eliminates repetitive measurements where errors commonly occur *(Figure 14-4)*.

To avoid confusion, the story pole should be cut the same height as the dividers, but it needs to be only 1 in. wide. Be sure to prominently label the bottom and top of the story pole. Lay out the bottom of the pole first and position the toe kick and bottom shelf. Next lay out the top—allowing for the crown molding (the total height of the ceiling) and the top rail (where the crown will seat). Finally, measure the remaining space between the top and the bottom shelves, and calculate the positions of the intervening shelves. Generally, I position taller shelves—about 16 in. high—near the bottom (for art books and such) and allow a minimum height of 10 in. for standard books.

Secure the pieces with a jig. With the story pole complete, it's a simple matter of transferring the layout marks to the dividers. But first, I make a

Grooves for Adjustable Standards

For many years, metal shelving standards were the most popular method for supporting adjustable shelves, and these products are still in common use. Though metal standards (now they're also manufactured in plastic) can be mounted on the surface of the case sides, I prefer to inlay these ugly beasts so that they're flush and less visible *(see photo)*.

Grooves for standards can be cut with a plunge router equipped with an accessory fence, or even faster, with a table-mounted router. A dado blade in a table saw is certainly the quickest method, but I always prefer using a router over a table saw because router bits cut much cleaner, with no tear-out, and the only sacrifice is a little speed.

To eliminate all that hole drilling, simply plow two 5/8-in. grooves, about 5/16 in. deep, and insert metal standards, and then use adjustable shelving clips to support the shelving.

Figure 14-4. *Use a story pole rather than a tape measure to lay out all joinery in cases, including bookshelves. A story pole provides a perfect full-scale opportunity to correct any math errors made in the drawing. And a story pole eliminates the possibility of measurement errors during layout.*

Figure 14-5. *The rails on the router jig are exactly the width of the router base, and they're waxed to ensure smooth travel. The center mid-span block, which straddles both case sides and holds them snugly together, can be moved to fit different widths of shelving.*

quick jig to secure all the pieces on the worktable. Since all the sides are the same length, I tack scrap material on my workbench so that all the pieces are trapped during layout and while routing the dadoes (*Figure 14-5*).

Dado both sides simultaneously. I like to dado both sides of a case simultaneously to be sure I have perfect pairs. I mark one side A-L and the other A-R (case A, left side and right side). Successive cases will be marked B-L and B-R, etc. I mark which end is up, too, which avoids simple errors during assembly.

As indicated earlier, I avoid cutting dadoes for the full thickness of the shelving. The thickness of plywood is always slightly less than the width of standard router bits, which results in a loose-fitting joint with little structural integrity or durability. Instead, for 3/4-in. shelving, I cut 1/2-in. dadoes in the case sides, and then use a bearing-guided rabbeting bit to trim the top face of the shelving so that it fits snugly in the dadoes. Cutting a rabbet that fits the dado precisely isn't difficult: I just adjust the depth of the router motor until the rabbeting bit removes just enough material to leave a tongue that fits in the dado tightly. The length of the rabbet's tongue is determined by the bearing on the bit (or on how long it has been since the rabbet has had a drink.)

Assemble the sides and shelves. Use bar clamps to be sure the shelves are completely seated in the dadoes (*Figure 14-6*). Temporarily secure the shelves with brads (the glue will need all the help

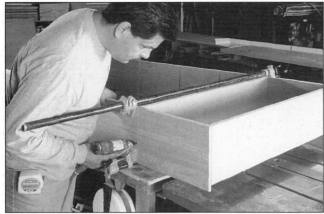

Figure 14-6. *Use bar clamps to squeeze the joints closed, and then nails to hold them until the glue dries. That way, the clamps can be removed right away.*

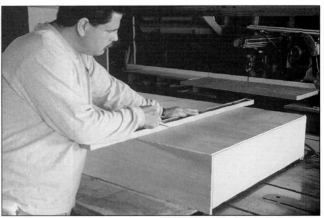

Figure 14-8. *Don't eyeball. Save time and the frustration of patching unwanted nail holes. Use a straightedge to trace a line down the middle of every shelf before nailing off the field.*

Figure 14-7. *Glue the back edge of every shelf, but place the back panel carefully — sliding it will smear glue everywhere. Position the edges flush with the ends before laying it down. Nail the first side, and use the back to square up the case.*

Figure 14-9. *Dado blades are great for making lots of molding designs, not just simple dados for shelving. This three-step nosing requires only two passes through the blade.*

it can get). In this instance, none of the nails will be visible because the finished sides (applied with screws from the inside later during installation) will cover the nail holes, and the other sides either meet stiles or end against walls.

Attach the back. Installing the back might seem a simple matter of spreading some glue and shooting a few nails, but there's a lot more to it. The back is an integral part of the construction: it squares up the case and provides both durability and rigidity. Start by gluing the edges of the shelves, and then carefully position the top and bottom edges of the back so that they're flush with the top and bottom of the case (*Figure 14-7*). Then drop the back onto the wet glue — without sliding it. Nail off that side, and then — before nailing the opposite side — rack the case so that the ends on the opposite side are flush, too. That way the case is sure to be square. A word of warning: Always check plywood for square before cutting it to size because factory sheets aren't always square.

Before nailing the field, carefully locate and draw lines across the back, centered above each shelf (*Figure 14-8*). That's the best way to avoid "shiners" (nails that miss the shelves).

Mill the nosing. After completing the cases, mill the stiles and the nosing. In this case I wanted a three-step nosing design, in keeping with the Arts and Crafts/Art Deco style of the bookcases. Cutting the steps was quick work with a table-saw-mounted dado blade (*Figure 14-9*). We attached the nosing with glue and brad nails.

Make the plinths and capitals. In this case blocks for the plinths and capitals were only marginally different from each other. Using the table saw, we chamfered both the top edge and the sides of the plinth blocks — in keeping with the Gothic look I was after. But for the capital blocks we cham-

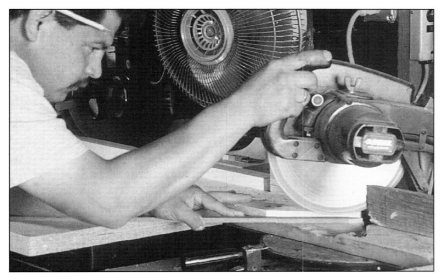

Figure 14-10. *An accessory fence is the only safe way to repetitively cut equally sized sharp miters.*

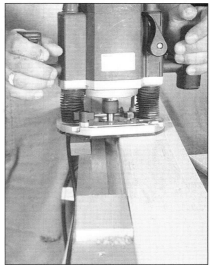

Figure 14-11. *Plow a ³/₄-in. groove up the middle of each stile. After this test piece, we installed a stop block to terminate each cut at exactly the right location.*

fered only the bottom edge, not the sides, because I planned to run crown molding around them.

As I said earlier, the top rail should be flush with the stiles, so that it's easier to run crown molding across the face of the case. However in this case, I wanted the crown molding to break forward around the capital blocks and create a deeper, more dramatic profile. But cutting the crown molding around the chamfers on the sides would have been a nightmare.

Adding simple inlays. I wanted to add another Gothic touch to the case, not only in the plinth blocks and rosettes, but in the stiles, too, so that they'd have the appearance of additional depth. But I didn't want the clutter of Gothic facets or steps in the stiles—that would be too much. Instead, I opted for a simple inlay. Yes, inlays can be simple.

Sharp arrow-like designs, whether in the shape of a cross, a sword, or even a fleur-de-lis, were common Celtic ornaments. To accentuate the stiles on this bookcase, I chose a simple walnut inlay with pointed ends.

First we ripped the inlays to exactly the width of my ³/₄-in. router bit—with which we'd later cut the grooves for the inlays. Then we cut the sharp miters using an accessory fence on my chop saw (***Figure 14-10***).

We could have used an accessory fence mounted on my plunge router to cut the grooves. Instead, we made a jig that ensured all the grooves would be wander-free. And with stops screwed in place, the jig also provided a means to terminate the cuts at precisely the same position (***Figure 14-11***).

The last step was the fun part. With the long arrow point overhanging the end of each groove, we held the inlays in place and used a utility knife to trace the miter onto the face of the stile (***Figure 14-12***). Using sharp chisels, we struck the outline of each miter, and then pared away the waste until the inlays fit perfectly (***Figure 14-13***). We spread glue in the grooves and snapped the inlays into place, sanding them after the glue dried (***Figure 14-14***). Because most of the job had been accomplished with a router and jig, the process wasn't nearly as time-consuming as it appears.

Installation

As always, if the drawings are precise and the construction is accurate, then the installation is always trouble-free. The cases in the next couple of photographs were designed to sit on top of three similar-sized file-drawer cabinets (there isn't enough room in this book to cover cabinets and drawers, too!), so setting the cases on top of the cabinet was similar to setting the cases on top of a separate toe kick (which is how I would have built this unit if it sat directly on the floor).

Attach the cases to each other. After stacking the three untrimmed cases on the cabinet top, I screwed the center meeting sides together, clamping them with the front edges flush (***Figure 14-15***).

BOOKSHELVES

Figure 14-12. *Pinch the walnut into the groove, and then, to ensure a tight-fitting inlay, scribe the outline of the pointed miter with a utility knife.*

Figure 14-14. *Glue the groove, bow the inlay to insert both ends of the miter, and snap the inlay into place. Then sand it flush.*

Figure 14-13. *Using only sharp chisels, strike inside the outline of the miter. Remove the waste, and then pare away the shoulders of the mortise until the inlay fits just right.*

Figure 14-15. *Clamp the meeting sides together. Use screws to secure them, but be sure the screws are not too long or they'll come out the other side.*

Figure 14-16. *Screw the finished side in place, too, but from the inside of the case, so the fasteners won't show.*

Then, using screws from the inside of the case, I also attached the finished side to the far case (*Figure 14-16*). I installed the trim (plinth blocks, stiles, and rosettes) after all the boxes were positioned and secured properly.

Trim the case. These units were designed to have a built-in appearance, yet—in the event of a furniture rearrangement or total relocation—remain somewhat portable. The installation of the plinth blocks and stiles provided for both requirements (*Figure 14-17*), but the crown molding really hid the truth.

To save costs (and scarce mahogany), I made the crown from several pieces of material. The face of the crown is a single piece of 1/2x2-in. mahogany, but the other two steps are only 1/2-x3/4-in. mahogany. The backing, like a Hollywood set, is fake—it's MDF.

As with most crown moldings, I preassembled as

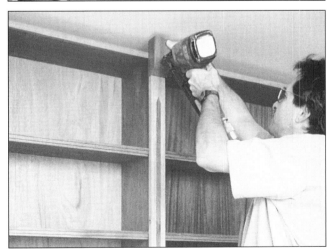

Figure 14-17. Install the plinth blocks first, using only a few nails (top). Next, attach the stiles, again, using only a few nails and no glue, if you might want to disassemble the unit (center). Then attach the capital blocks (bottom).

Figure 14-18. Crown molding can be built up from several pieces, and the backing doesn't have to be stain-grade (top). Preassemble every possible joint in the crown molding, using spring clamps to hold the miters tight while brad-nailing (above).

Figure 14-19. Install the crown molding all the way to the ceiling. Using a wider reveal on the top rail of the bookcase will allow more wiggle room and won't telegraph inconsistencies in the ceiling or the crown.

Figure 14-20. *Though it is short of the ceiling, the crown molding on these traditional bookcases highlights the sloping ceiling.*

many pieces as possible, using miter clamps, glue, and brads (*Figure 14-18*). Then I installed the completed long sections with as few nails as possible, so that it would be easy to remove if the time ever came to pack up the books and move (*Figure 14-19*).

A TRADITIONAL BOOKCASE

Most bookcase construction is pretty much the same (except for the differences between adjustable and stationary shelves, but more on that in a moment). It's the trim that makes each job stand out. Traditional bookcases have standard stile and rail sizes, usually about 2 in. wide. But to enliven the look, crown molding (even if the cases don't extend all the way to the ceiling) and wall paneling should be added whenever possible. After all, these additions spell the difference between standard factory fare and custom-designed, handcrafted woodwork.

Layout

On one job we were asked to build built-in cases around a bay window. The room had a tray ceiling (like a single, sloping coffer) so we designed two flanking units connected by a center shelf (*Figure 14-20*). To complement and complete the traditional design, we added a short panel of wainscoting beneath the bay window.

Bookshelves, like furniture, tend to close in a room. To maintain the greatest appearance of space, I designed these shelves to be only 10 in. deep. At that narrow depth, the additional shadow line of a toe kick can look silly. Instead, I elevated the lowest shelf 2 in. above the floor, to provide a safe clearance between books and vacuum cleaners or toes. However, I left the face flush to match the top and also to avoid a toe-catcher.

The height of the cases are determined by the height of the crown molding and the distance the

Figure 14-22. *From a distance, it might seem that the cases are shoved against the casing and fastened to the wall, but there's more to the installation than that.*

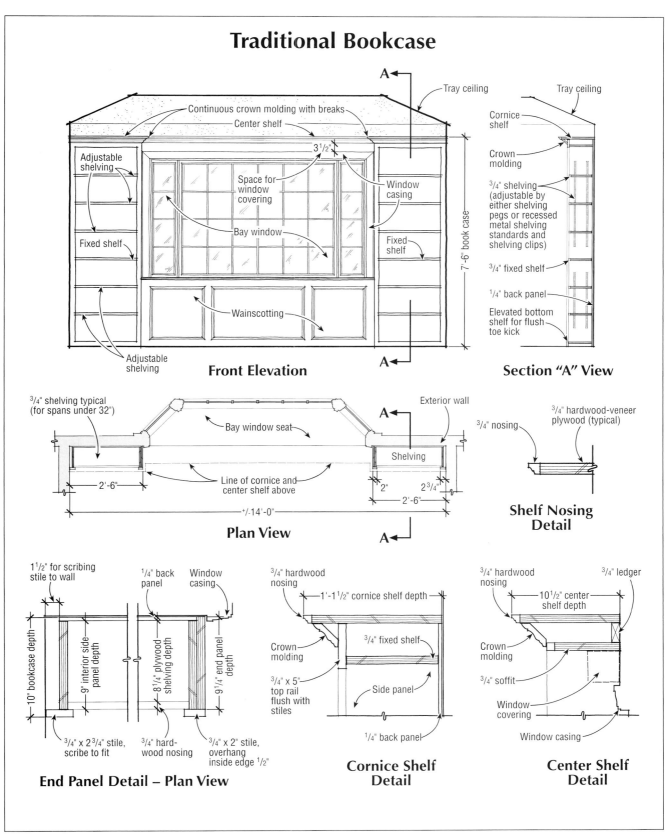

Figure 14-21. Crown moldings and paneled wainscoting dress up these custom shelves built around a bay window with a single shelf above the window. The height of the cases is determined by the distance required for window treatments between the window head casing and the bottom of the center shelf. Fixed shelves at center and bottom strengthen the cases, with the bottom shelf also serving as a toe kick. The end stiles overhang the end panels for scribing as shown in the End Panel Detail.

Figure 14-23. *To locate cutouts for electrical outlets, first establish a plumb line at the exact location of the case.*

Figure 14-24. *Measure electrical outlets from the plumb line, and then make cutouts from the back of the case, to avoid tear-out.*

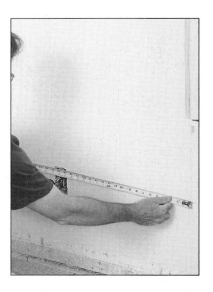

owner wishes to maintain between the bottom of the center shelf and the top of the window casing—a space reserved for drapes or other window covering.

With the bay window cases, I started, as always, with a sketch of the unit, and then, with the crown molding in hand, I made a more careful drawing (*Figure 14-21*). Then I assembled the rough cases in my shop.

Installation

After the cases are built, it might seem that all that's required is slapping them into place and screwing them to the walls. But there's much more to the installation than that.

Layout for cutouts. I start by determining exactly where I want the cases to be secured to the wall. In the room in *Figure 14-22*, I wanted them tight against the casing, but the casing and the window were out of plumb. Rather than install the bookshelves out of plumb, I removed the casing and reinstalled it, splitting the difference in the reveal on the jamb. The casing was still slightly out of plumb, so I decided to hold the bookcases tight against the casing on the bottom and 1/8 in. away from the casing at the top, and then caulk the gap when the installation was complete. If the work had been stain-grade, I would have scribed a plumb line on the back of the casing and then ripped it before installing the bookcases. But even before I could install these paint-grade cases, I had

Figure 14-25. *A Kreg Mini Jig® is great for fastening cabinets and bookcases to walls. Find the exact center of the studs, and then use a drywall screw long enough to penetrate through the case, through the drywall, and at least 1 in. into the stud.*

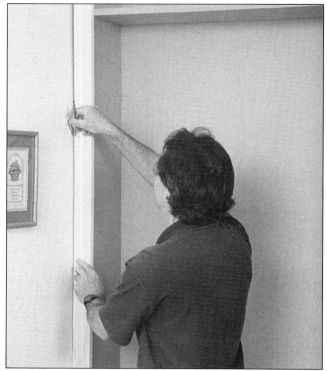

Figure 14-26. *Temporarily tack the stile in place so that it's parallel with the case, and then scribe the contour of the wall onto the outside edge of the stile. Spread the scribes so the remaining stile will equal the opposite stile.*

Figure 14-27. *Use a mockup — not math — to find the exact location where the cornice shelf must break back.*

Figure 14-28. *Place the two end sections of the cornice shelf on the cases, and then measure for the center narrower section over the bay window.*

to make cutouts for the electrical outlets.

Using a long level to represent the position of the bookcase, I traced a line down the wall (*Figure 14-23*). Measuring from that line, I located the outer dimensions of the electrical outlet, and then measured from the floor to find the height (*Figure 14-24*). Then, to prevent tear-out and to ensure tight-fitting outlets, I made the cuts from the back of the cases.

Fasten the cases and scribe the stiles. I frequently drill pocket screw holes using my Kreg Mini Jig® (Appendix 1), especially for mounting cases to walls. Though the bit can be used freehand without the little jig, it's safer and more accurate to use the jig. I find the exact location of the studs, and then drill a pocket hole in the top of the case. It takes only two long drywall screws to secure a bookcase (*Figure 14-25*).

Once the case is fastened to the wall, I start installing the trim, beginning with the stile against the wall. I position the stile so that it's parallel with the case, and then tack it in place with a brad or two. Sometimes walls are terribly out of plumb or bowed, and I always arrive armed for that possibility — for stiles that may need to be scribed, I mill the stock in my shop an extra $3/4$ in. wide, just in case. But I hope to scribe enough off the outside edge of the stile so that the finished width will be a close match with the opposite stile (*Figure 14-26*).

The stiles on the wall side of each case must be scribed before they're permanently fastened to the case, but the stiles on the window side of both cases are simply nailed flush with the outside of the case. And then the top rail and bottom rails are installed. The top shelf inside the case is located several inches below the top of the case to allow for the crown molding. And the top rail is installed $1/2$ in. lower than the inside of the top shelf, so that the overhang is similar to that of the stiles.

Cut the cornice and crown. The cornice shelf cantilevers over the top of each case and spans the distance between them, providing a ledge for the crown molding that also runs from bookcase to bookcase. After the crown molding is installed, the cornice shelf will still project a little past the crown molding — for a reveal. In this case, I wanted the cornice-shelf reveal to be about $3/8$ in. That's pretty simple.

However, because the crown molding and the cornice shelf both break back at each end of the bookcases, and because I wanted the cornice-shelf reveal to be an even $3/8$ in., I was faced with some confusing math: the projection of the nosing on the shelf, the projection of the crown molding from the side of the bookcase, and the amount of the reveal.

Instead of risking a math error — even with a calculator — I simply mocked up a piece of the cornice shelf and the crown molding, and then had my co-carpenter mark the location of the break, right on the ledger (*Figure 14-27*).

He first cut the parts of the cornice shelf for each bookcase, put them in place, and measured the length for the shelf between (*Figure 14-28*). Next we kerfed the ends of all three pieces with a $1/4$-in. slot

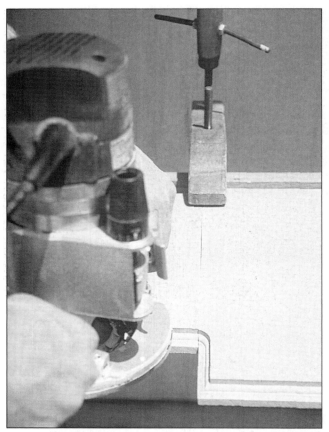

Figure 14-29. *First cut slots and temporarily install splines in the joints, and then clamp the end shelves to the center shelf and run a beading bit on the edge.*

cutter, installed splines (without glue, see Chapter 11, "Decorative Walls" and Chapter 12, "Decorative Ceilings"), clamped the pieces to my chop saw extension wings, and ran a bearing-guided beading bit around the front edges (**Figure 14-29**).

Install the cornice shelf and crown. Because the combined cornice shelf was long, heavy, and tight-fitting (and a little wiggly, with a joint at each end), we installed each piece in place, bowing up the last two pieces so that the spline (now glued) would snap into the kerf (**Figure 14-30**). We installed a second shelf—or soffit—beneath the upper shelf, to provide backing for the bottom of the crown molding as well as backing for window coverings (**Figure 14-31**).

As I've said before, crown molding is easier to install if it's preassembled. We cut the crown for this unit and preassembled the outside miters that wrapped around each bookcase. Before installing the ends, we tacked in the long piece between the cases, leaving the corners loose. Then we installed the preassembled pieces, matching the inside miters perfectly (**Figure 14-32**).

SUSPENDED BOOKSHELVES

There's one other type of bookshelf arrangement that should be discussed in this chapter, and that's suspended shelving—bookshelves or cases hung from walls.

Standard case construction is often used for suspended shelving, with the sides acting like corbels

Figure 14-30. *Glue up the slots and splines, then bow the center shelf up to install the last end shelf.*

Figure 14-31. *Install a soffit beneath the center cornice shelf—as backing for the crown. The bottom of this soffit must be installed at a distance beneath the bottom of the cornice shelf that is equal to the rise of the crown molding. The hollow space between the two shelves provides a perfect place for a ledger on the wall, which secures the long span.*

Figure 14-32. *To speed production and ensure tighter joints and straighter molding, preassemble the crown, especially the outside corners.*

or brackets. For additional security, cleats are usually added beneath the shelves, so that the unit can be fastened to the wall — without fear that only a flimsy back separates hundreds of pounds of books and MDF shelving from a homeowner's head.

But on the job in **Figure 14-33**, the homeowner wanted the shelving to mimic the span of the office desks — without interfering with the open-beam look of the room. For that reason, we used a different type of material for the shelving—2x12 #2 pine.

Layout

With only two shelves, I didn't even consider making them adjustable. Besides, I wanted to build this

Figure 14-33. *The long span of these bookshelves required heavier 2x12 lumber. But the two pairs of doors break up what would have been a boring set of shelves.*

Figure 14-34. *Use a story pole even for simple bookshelves.*

Figure 14-35. *To ensure straight and square dadoes, transfer the layout lines across the sides.*

Figure 14-36. *Because of the offset between the outside of the template guide and the cutting edge of the router bit, the template will be wider than the layout lines, so center the template over the lines.*

Figure 14-37. *Use a plunge router and make several passes to produce a 1/2-in.-deep dado that accommodates the full thickness of the 1 1/2-in.-thick shelving — no need to rabbet the shelves.*

Figure 14-38. *Three-inch drywall screws — and glue— ensured that these shelves wouldn't sag.*

dado. This thick material was plenty strong enough to span the slightly more than 3 1/2-ft. distances between dividers, so I broke the overall length into four nearly-equal sections — following the layout of the desk units below.

Even though there were only two shelves, I still made a story pole (**Figure 14-34**). I didn't want to risk having one shelf slightly off-center from another flanking shelf. I used the story pole to lay out the dadoes in each divider, and a framing square to transfer the layout lines across the entire face (**Figure 14-35**). On the center dividers I extended the layout lines around the front edge and across the back face, too, so that they were visible on both sides. For the mid-span dadoes on the shelves, I simply sandwiched the shelves together and traced the layout with a framing square. I

configuration like a tank, so it would never settle or move. Because the overall span was almost 16 ft., I designed the shelves in two separate units, and planned to connect them on the job at a common

Figure 14-39. *I fastened the dividers with screws driven straight through the shelving.*

Figure 14-40. *The upper book rests were secured with screws driven through pocket holes — drilled with a Kreg Jig®.*

decided the dadoes would be ³/₈ in. deep, leaving ³/₄ in. of wood at the mid-span dividers.

Construction

Because the material was 1¹/₂ in. thick, I didn't have to worry about dadoing the sides *and* rabbeting the shelves. Instead, I made a quick jig for my plunge router. The template guide for my ³/₄-in. router bit is 1 in., so I made the jig with a 1³/₄-in. opening and tested it before going to work on all the dividers. After placing the jig on the material, I tapped it until the layout lines were perfectly centered between the jig rails (***Figure 14-36***), and then made each cut in two or three passes (***Figure 14-37***).

Assembling this beast was a bit out of the ordinary. I glued every joint and used 3-in. drywall screws to make each one permanent (***Figure 14-38***).

I was able to screw the mid-span dividers from the top and the bottom, but not the short blocks that I had included at the last minute to serve as upper book rests (***Figure 14-39***). (It dawned on me that extending the mid-span dividers would make great rests for books on the top shelf, and then I realized I could have made the mid-span dividers the same height as the ends and cut the shelving instead! That route would have been a little simpler.) So to install the short blocks, I drilled them with pocket holes (***Figure 14-40***), which I later filled with handy plugs made just for that purpose by the Kreg Tool Company (Appendix 1).

Installation

Holding these heavy shelves on the wall while simultaneously running screws into the studs was out of the question. Instead, a co-carpenter and I used a laser to quickly mark a level line across the wall. Next, we found the center of each and every stud. Then we fastened a temporary cleat to the wall, just beneath the line of the shelving (***Figure 14-41***). I knew we'd be leaving a few holes to patch, but the shelves still had to be painted anyway, and besides, I'd rather leave a few holes in the wall than a crushed computer on someone's desk.

We hoisted the first unit on the cleat, and while I held it another carpenter used the Kreg Mini Jig® to drill pocket holes at each stud, and

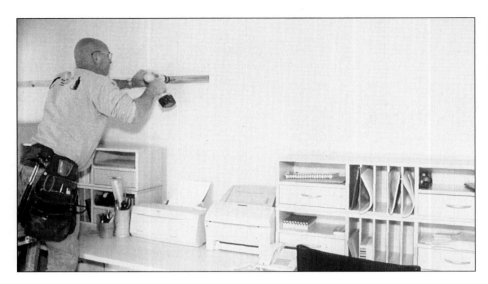

Figure 14-41. *A temporary ledger on the wall simplified the installation and reduced the risk of an accident while working above a desk and computers.*

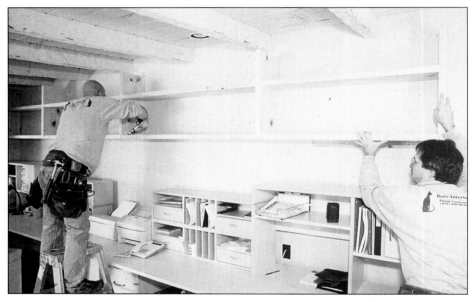

Figure 14-42. *Before attaching the second unit to the wall, we glued the dadoes and drew it tightly against the first section, and then secured the two with more screws.*

then drove 3-in. drywall screws into every one. We repeated the same sequence for the second unit, though once we had it against the wall on the cleat, we joined the two sections first (*Figure 14-42*), before driving fasteners into the wall studs.

To break up the boring look of such long shelves, we scribed simple doors—glued up from poplar—into the two end openings, and made wooden pulls for each door. We used a simple block as a jig so that the pairs of pulls were evenly located (*Figure 14-43*). Less than five minutes after this last picture was shot, the owners had the shelves filled with books. Too bad they had to remove everything a week later for the painter.

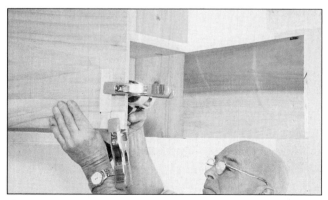

Figure 14-43. *We scribed simple doors to fit the two end openings and hung them on butt hinges. Then we used a simple spacer block to make sure that the wooden pulls were attached symmetrically.*

Chapter 15

Mantelpieces

This is the last chapter of my book and if you haven't guessed already, I'll come right to the point: I enjoy finish carpentry. And I've saved my favorite subject for last.

Making mantelpieces is a fitting last chapter for this book because a beautiful mantelpiece requires a combination of skills. Whether it's a small, simple, and inexpensive design or a large, elaborate, and costly architectural statement, mantelpieces are miniature edifices that often require every inch of carpentry know-how that I've covered in this book—from baseboard to casing, from crown molding to panel molding, from table-saw joinery to routing flutes, rabbets, and beaded edges *(Figure 15-1)*.

But making a mantelpiece depends on more than a carpenter's skill at cutting and applying

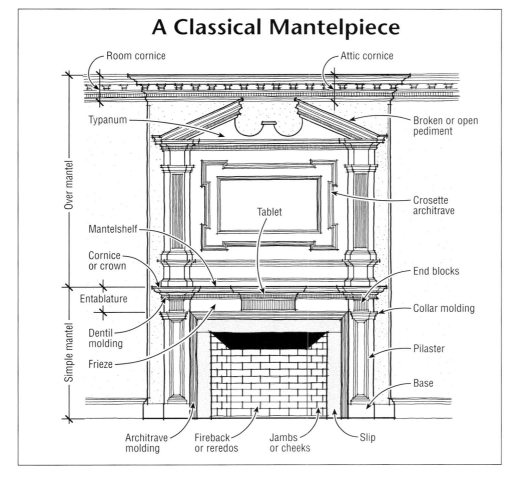

Figure 15-1. *Classical mantelpiece designs originate from Greek and Roman temples, but they are also influenced by Renaissance inventions. Broken entablatures and pediments were introduced during the late Renaissance Baroque period.*

Figure 15-2. *Pilasters are easy to make using hollow-box construction in almost any sheet material. Mitering the outside corners produces the look of solid stock (left). Glue the miters, carefully wipe the excess glue off while it's still wet, and then gently ease the edges with a block of wood (right).*

Figure 15-3. *Use a homemade miter box to cut huge moldings. Stop blocks inside the miter box hold the material securely in position.*

molding. Recall the story from Chapter 1 about those two "fanciful" mantelpieces and you'll know what I mean: Mantelpieces—like most of finish carpentry—are all about design.

However, there isn't enough space in this book to exhaust the intricacies of mantelpiece design. Instead, my aim here is to explain the basics of mantelpiece construction and shed some light on a few challenging installations. Along the way I'll explain the reasons for the details and the proportions that I use.

DESIGN & CONSTRUCTION

The two principal parts of a mantelpiece, the pilasters (or columns) and the entablature (frieze, crown moldings, and mantelshelf), have been around for centuries. While the basic form is taken from classical Greek temples, such as the Parthenon and the Erechtheion, the style most familiar to us is Baroque and originated during the late Italian Renaissance.

Italian architects, especially Michelangelo, added a little extra detail and shadowing to the Greek orders by breaking the pilaster through the entablature (sometimes breaking through only the frieze, sometimes breaking through the frieze and the cornice). The drama of a contemporary mantelpiece depends on the same technique.

And although mantelpieces are divided into two main architectural elements, as a carpenter I find it easier to approach most mantelpieces in terms of three primary pieces: the pilasters, the frieze, and the cornice.

The Pilasters

Pilasters are the place to start because they support the mantelshelf, and in many cases the cornice can't be built until the pilasters are assembled. I always build pilasters tall enough to reach past the frieze and support the mantelshelf. Which brings me to the height of the pilasters.

The cornice, the type of frieze, and the design of

the end blocks should determine the height of the mantelshelf, but often it's the height of the fireplace opening and the look of the room that determines both the frieze and the end-block design.

In this case *(Figure 15-2)*, I made the mantelshelf 57 in. from the floor; any higher and the thing would have been towering beside a standard 6-ft. 8-in. door. I knew I'd be making the mantelshelf from two pieces of 3/4-in. MDF, so to determine the height of the pilasters, I subtracted 1 1/2 in. from the overall height of the mantel.

Experience has taught me that, at that height, the best width for the pilaster is between 5 in. and 8 in.—on the wider side for larger rooms and Georgian styles, but toward the narrower dimension for smaller rooms and lighter Neoclassical styles.

I made these pilasters 6 in. wide. If the pilasters are less than 1 1/2 in. deep, then I sandwich two layers of material together. But, for the most part, I build pilasters that are between 2 in. and 6 in. deep, so I make them in the form of a hollow box, with mitered corners (though I often make Craftsman-style pilasters in solid wood with butt joints).

Laying out and assembling the pilasters—which includes the end blocks—is one third of the fun of making a mantelpiece. But the task can be confusing. I try to simplify the concept by thinking of the pilaster as two parts—the upper pilaster (consisting of the end block and sometimes the cornice) and the lower pilaster (the fluting, paneling, or other design on the lower pilaster leg).

First, the upper pilaster. As I said earlier, the height of the cornice and the height of the frieze influence the layout of the upper pilaster. Of course on real wood-burning fireplaces, the height of the frieze is also limited by building code restrictions, which often require a 12-in. noncombustible area above the fireplace opening (if you're designing a mantelpiece, always check with your local codes to be sure of this dimension).

Rather than working with intangible measurements and numbers, I prefer going right to work with the moldings. I start laying out the pilaster by assembling the cornice that will support the mantelshelf—in this case a piece of Fypon molding mitered to run around the pilaster legs, creating the dramatic "break" seen in many Georgian mantelshelves *(Figures 15-3 to 15-6)*.

To determine the length of the return pieces on the inside of the pilaster legs, I use a short piece of the frieze as a gauge block. On this mantelpiece, the frieze is another single piece of polyurethane

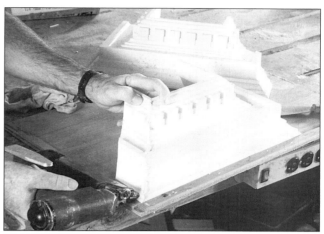

Figure 15-4. *Preassemble the outside corners of the cornice molding. An assembly board ensures that every corner will be perfectly square. For urethane moldings, use plenty of the manufacturer's recommended adhesive—and plenty of solvent to clean off the excess.*

Figure 15-5. *Start the pilaster layout by locating the exact position of the cornice molding at the top of the pilaster. The top of the molding should be flush with the top of the pilaster and provide support for the mantelshelf.*

Figure 15-6. *If necessary, temporarily fasten the cornice to the pilaster so that it's perfectly square to the pilaster.*

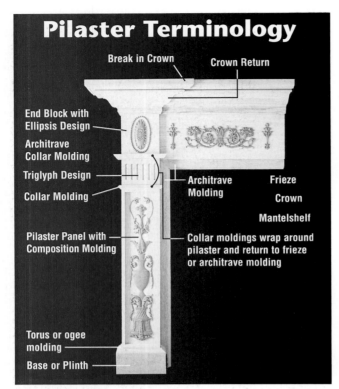

Figure 15-7. *Pilasters not only support a mantelshelf, they frame the fireplace opening. Attractive pilasters can be fashioned with few moldings and simple designs.*

Figure 15-8. *Determine the height of the frieze (here I used a small scrap of the frieze material), and then install one side of the face frame panel so the top of the panel molding will be flush with the bottom of the frieze.*

Figure 15-9. *Rip and resaw the fluting to fit between the sides of the face frame. The height should be between 3 in. and 5 in.*

Figure 15-10. *Fasten the fluting and install the mid-rail of the face frame.*

molding, but countless designs are possible (more on frieze designs in a moment).

I tack the crown molding temporarily in place, hold the piece of frieze at the bottom of crown, and mark the bottom of the frieze on the pilaster leg. For the most part, I try to match the bottom of the end block—usually distinguished by a collar or ogee molding—with the bottom of the frieze, which is the location of the architrave molding on a classical order. But like some rules, that one is often ignored *(Figure 15-7)*.

Then, the lower pilaster. Once the bottom of the end block is located, I begin work on the lower pilaster. This mantelpiece is a combination of Georgian and Federal details. The cornice breaks around the end block, and the mantelshelf will have Georgian "wings" that correspond with the breaking cornice. However, the pilasters will be more in the Federal style: slender and decorated with flutes and panels, rather than with heavy brackets or corbels.

I start by installing a 3/4-in. stile-and-rail outline for the flutes and panels (the flutes are meant to look like the triglyphs on the Doric order (see

Figure 15-11. *Next, work from the bottom. Install backing for the base molding, and then install the bottom rail. Make the bottom of the rail flush with where the top of the base molding will be.*

Figure 15-12. *Install the base molding now or later.*

Figure 15-13. *Use a jig to cut the panel molding and spring clamps to preassemble the panels.*

Figure 15-14. *Fasten the panel with glue and nails.*

Figure 15-15. *Reverse the panel molding to create a decorative raised panel.*

Chapter 1, Figure 1-1, page 3). Triglyphs are always placed directly above each column, and they alternate with blank spaces (metopes) across the remainder of the frieze. Relying on photographs of Federal-style mantelpieces, I made the short-fluted section almost 5 in. tall. Though I've seen the detail as short as 3 in., I wanted the extended length of the fluting to complement the height of this mantelpiece *(Figures 15-8 to 15-10)*.

After finishing the triglyph section, I turn to the lowest part of the pilaster. In some cases, the base detail is a continuation of the baseboard in the room. However, in the example shown in *Figures 15-11 and 15-12*, the base is a distinctly different profile, made taller in order to provide a point of termination for the baseboard in the room.

After finishing the base, all that remains is the panel and the end-block decoration. For the panel, I chose the same panel molding used in the room's wainscoting—stain-grade mahogany for the mantelpiece and paint-grade poplar for the wainscoting *(Figures 15-13 and 15-14)*.

Elaborating the end blocks. I could have used any number of devices to decorate the end blocks, but after playing with several different ideas, I decided to use the same panel molding again, except on the end blocks I reversed the molding to form a rectangular, raised rosette *(Figure 15-15)*.

To add more intensity to the rosettes, I made

Figure 15-16. For the center panel, use a contrasting wood. Walnut works well with mahogany. Rip and resaw the stock to size.

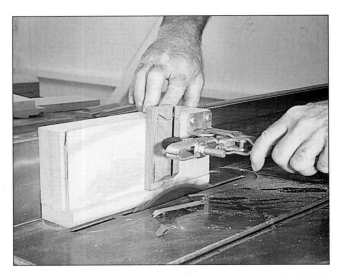

Figure 15-17. With a straight-cut bit and router jig (always use a jig to keep your fingers safely away from the bit), cut a raised rectangle in relief in the center of the panel.

Figure 15-18. Chamfer the edges of the raised rectangle with a table saw. Use a push block to keep your fingers away from the blade (top). Insert the panel in the panel frame (middle). Then fasten the frame and panel to the end block with glue and brads (bottom).

the center insert from walnut *(Figure 15-16)*. My first attempt at an insert was a simple flat mini-panel, but I didn't find that especially pleasing. Next, on the same blank panel, I used a router bit to cut in relief a long raised rectangle on the panel *(Figure 15-17)*. But that still wasn't exciting enough for me. On the third try, and still working with the same small piece of wood, I cut a chamfer around the center-raised rectangle. Now that pleased me just fine *(Figure 15-18)*. After installing the rosette and panel, I finished the end block by applying an ogee collar molding at the top of the fluting *(Figure 15-19)*.

The Frieze

Though the frieze in this mantelpiece is made from a single piece of molding *(Figure 15-20)*. I usually make the frieze from another U-shaped hollow box

Cutting Custom Fluting

You can't always get what you want when it comes to fluted casings or panels, especially if you prefer the look of stopped flutes to common commercial profiles. Stopped flutes always seem to have been carved by hand, even when they're cut with a router.

I use several methods for cutting flutes, but the easiest and fastest way I've found is with a table-mounted router. I prefer a 1/2-in. roundnose (core box) bit, set to a depth of 1/4 in. To make the process easy, I rip some MDF spacer strips the length of the fluted column. For the fluting on these columns I make the strips 1 in. wide: 1/2 in. for the flute plus 1/2 in. for the space between flutes.

I start by placing the spacer strips between my rip fence and the work piece, and then remove one strip after each pass so it isn't necessary to move my rip fence for each of the flutes. This makes it easy to make more panels later, if I need to.

To stop the flutes 1 1/4 in. from the end of the board, I draw a line across all the 1-in.-wide strips perpendicular to my rip fence and exactly 1 1/4 in. beyond the fluting bit. This line locates the beginning of the fluting. I carefully lower each piece ahead of this line (toward the outfeed table) for the initial plunge, and then slowly move the workpiece back to the line (pulling the piece toward me) so that all the flutes begin 1 1/4 in. from the top of the board *(see photo)*. To make sure all the flutes end 1 1/4 in. from the bottom of each board — and to hold all the spacer strips in place — I screw a long stop block to my outfeed table.

After fluting the faces, I rip each edge at a 45-degree bevel to prepare for assembling the pilaster boxes.

Like many tricks of our trade, it's difficult to accomplish the routing procedure with a guard in place, so it can be dangerous. Though the bit is covered by the stock at all times and does not cut through the material, I take special care to keep my hands away from the cutting area.

Another way to cut fluting is to run a router along a straightedge or a router guide. While these methods might be safer, they are much slower and less accurate than using a table-mounted router.

Use a router table to flute the pilaster faces. Register the work piece against 1-in. strips of MDF, rather than against the fence. After each pass, remove one strip before making the next flute.

Figure 15-19. *Finish the pilaster by applying collar molding at the top of the fluting, which separates the end block from the body of the pilaster.*

of MDF or stain-grade MDF-core plywood. Then I install the frieze between the pilaster legs, as in *Figure 15-21* (I'll cover installation more completely later).

For just a moment I'll leave the mantel I've been working on—with it's simple piece of frieze molding—in order to illustrate on another mantel a few important points about frieze designs as well as frieze and pilaster assembly.

A flush frieze. The frieze can be built at exactly the same depth as the pilasters so that it's flush with the face of the pilasters. In that case, the cornice molding won't break around the pilasters but will instead run in a continuous straight line from one end of the mantelshelf to the other *(Figure 15-22)*. That design is a good alternative for many rooms where Georgian-style overstatement wouldn't be

Figure 15-20. *This frieze is cut from a single piece of urethane molding, decorated with urns, anthemions, and garlands.*

Figure 15-21. *Most often I install a third U-shaped box constructed for the frieze and cut to fit snugly between the pilasters.*

in keeping with the furnishings or the size of the room—mantelpieces with dramatic breaks aren't comfortable in most traditional homes.

A frieze flush with pilaster backboards. The frieze can also be designed so that it's flush with the backboards that are sometimes installed behind the pilaster legs. Then the entire cornice and cornice soffit can rest on top of the pilasters, which also eliminates the intricacy of the crown molding breaking around the pilasters *(Figure 15-23)*. Or, for a different look, the cornice can be lowered onto the pilasters, in which case the cornice molding alone, and not the mantelshelf, will break forward around the pilasters *(Figure 15-24)*.

Elaborate Hollow Pilasters

A pilaster can also be built in several pieces, rather than as one hollow box. I knew this mantelpiece would be prefinished—stained and lacquered before installation—and I also guessed that during the installation I'd have to scribe the pilasters to the hearth and maybe even to the soffit under the mantel. So I decided to make the entire assembly from separate boxes *(Figure A)*. That way, when it came time to scribe the pilasters to fit, I had only to cut through one layer of 3/4-in. material. To help see the scribe line, I applied masking tape all around the top of each pilaster *(Figure B)*.

Figure A. *Pilasters can be assembled in pieces, too, especially if they must be scribed to fit. If the top and bottom details require several thick steps or reveals, then the pilaster can become too thick to cut with a fine-bladed saw.*

Figure B. *Apply masking tape to prefinished woodwork so that scribe lines show up clearly, even while the sawdust is flying.*

Figure 15-22. *A straight cornice (crown and mantelshelf) blends better with traditional furnishings and doesn't demand too much attention.*

Figure 15-24. *In this mantelpiece, the crown molding breaks slightly around the pilaster, though the mantelshelf does not.*

Figure 15-23. *The frieze in this example is flush with the pilaster backboard. A heavy collar molding (part of the lower cornice) breaks around the pilaster, but the primary crown molding beneath the mantelshelf does not.*

Figure 15-25. *A deep frieze is framed by prominent pilasters, accentuated here by the cornice (the crown molding and the mantelshelf) breaking around the pilasters. The composition molding applied to this mantelpiece imitates an early 19th-century Adam-style mantel.*

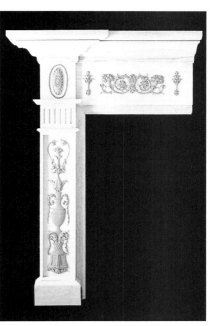

A deep frieze. Like most finish carpenters, I favor the two additional outside corners created by the break in the crown molding. I also prefer a deep break—one that's at least 1 1/2 in. deep—so that it's easier to work with the small pieces of crown molding. Whenever possible, I build the frieze at least 1 1/2 in. shallower than the pilasters—so much for classical proportions *(Figure 15-25).*

Decorating the frieze. Mantelpieces are like fingerprints—no two need ever be the same. There are countless ways to build and decorate the pilasters, the cornice, the end blocks, and even more methods and materials with which to design and decorate the frieze.

On the Doric order, triglyphs were separated by blank spaces (metopes). Those black spaces were decorated with Bull heads, or rosettes (dogwood blossoms were popular in Colonial America), or mythological figures. The Corinthian frieze, on the other hand, was a continuous mythological scene, without triglyphs or metopes.

During the early 19th century, composition molding was often applied to the frieze, in all manner of designs, from garlands and anthemion stems, to medallions picturing mythological scenes (see Figure 15-25). I'll note throughout the remainder of

MANTELPIECES 245

Figure 15-26. *Lay the crown molding on the bottom of the mantelshelf and mark the footprint of the crown—making sure the overhang, or reveal, is equal on the ends, the front, and the sides.*

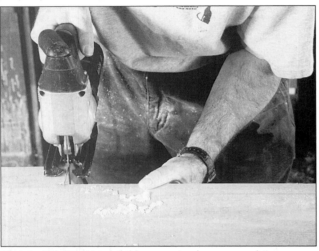

Figure 15-27. *Use a table saw, circular saw, and jigsaw—whatever you need to make perfectly straight cuts.*

this chapter several different methods used to design and decorate the frieze, though the possibilities are unlimited.

The Cornice

Mantelshelves are deceiving. Novice carpenters think that all the effort on a mantelpiece goes into the crown molding and the mantelshelf. Nothing could be farther from the truth. As I've shown, the pilasters can be far more complicated. However, building a mantelshelf with wings is no simple task. I start by building the entire cornice, and then I gauge the dimensions of the mantelshelf (the overall length and width—including the cantilever around the cornice wings) from the finished cornice, rather than relying on math.

The cornice wings were built around the end blocks before laying out the pilasters, so I needed to

Figure 15-28. *Fasten the nosing to the shelf with glue and nails.*

Figure 15-29. Then fasten the crown to the bottom of the mantelshelf.

Figure 15-30. The paneling in the pilasters matches the profile in the wainscoting. Notice also that the collar molding terminates just above the chair rail, which avoids a molding conflict.

cut only one more long piece of cornice, from the inside corner of one pilaster to the inside corner of the other.

But before I could do that, I needed to determine the width of the opening between the pilasters. Because this mantelpiece was purely decorative (built for JLC LIVE, Las Vegas, 2000), I decided to spread the pilasters apart the distance of the fireplace opening (36 in.) plus 6 in. on each side of the fireplace to allow for simulated slate (12 in.). So I cut the long piece of cornice 48 in. from long point to long point. I repeat, be sure to consult local building codes to determine the proximity of combustible material to a fireplace.

Next, I assembled the cornice completely, then turned it upside down on top of the material I had sandwiched for the mantelshelf (two pieces of mahogany plywood with an MDF core). From that point, it was easy to measure the outline of the cornice onto the bottom of the mantelshelf and maintain equal overhangs in all directions. I made the overhang only $3/8$ in., because I knew that the mantelshelf nosing would add an additional $1 1/4$ in. (*Figures 15-26* to *15-29*).

INSTALLATION

Mantelpiece installations vary, too, but fortunately not as much as mantelpiece design. The most straightforward is a blank wall, where the pieces are simply fastened in place — first the pilaster legs, next the frieze, then the cornice, and finally the trim *(Figure 15-30)*. But carpenters encounter other challenges in the field, too, and none seems more demanding than building a mantel and pilasters around an existing hearth or brick sur-

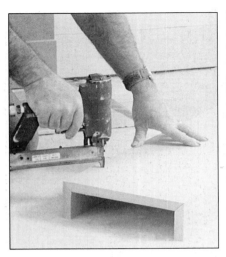

Figure 15-31. Hollow-box construction entails mitered corners and just about any sheet material, and it also provides for an easy installation.

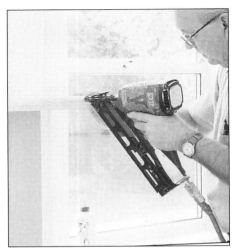

Figure 15-34. Laminate the mantelshelf from two pieces of MDF, scribe the shelf to fit the back wall, and then fasten it to the wall and the pilaster/frieze backing. Afterward, apply the nosing.

Figure 15-32. Cut the backing block to fit (subtract 1/16 in. to be sure) inside the hollow pilaster. Then fasten the backing to the wall with adhesive caulk and nails — find the studs first with a long finish nail, or if there aren't any studs, use plastic plugs and screws to secure the backing until the adhesive dries.

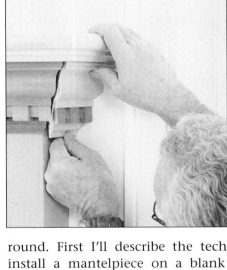

Figure 15-35. The crown molding can be preassembled or installed in pieces right on the mantel. I prefer preassembly because the results are better.

Figure 15-33. A hollow-box frieze is installed the same way: first the backing, then the box. Nail through the top and bottom of the frieze into the backing.

round. First I'll describe the technique I use to install a mantelpiece on a blank wall, then I'll cover two ways for tackling mantelpiece installations over existing brick surrounds.

A Flat-Wall Installation

Installing a mantelpiece on a flat wall is the most common situation I encounter, especially on new construction. Start by locating plumb and level lines on the wall — both the inside and the outside edges of the pilasters and the mantelshelf. If the pilasters are flat or composed with backboards, secure them directly to the wall with adhesive caulk and nails, but first find the studs with a 16d finish nail. Whether the fireplace is metal or masonry, a stud is almost always located at the edge of the fireplace box. However, if you can't find a stud, then use plastic anchors, screws, and adhesive caulk to secure the pilaster to the wall.

The pilasters. If the pilasters are hollow boxes, I start by installing a backing board on the wall (**Figures 15-31** and **15-32**). Rip the backing board

Starting at the Top

Rules ensure safety, speed, and accuracy, but from time to time some rules need to be broken. I usually install every mantelpiece from the bottom up, starting with the pilasters. But after setting up all our tools to install this prefinished mantelpiece, we discovered that the pilasters hadn't been stained the same color! Back to the finishers they went. I was determined to still get something done. But I didn't want to take the mantelshelf with me or leave it loose on the job—even my best friend might have taken this one.

So I started from the top *(Figure A)*. I screwed to the wall a ledger that had been precut to exactly the right width and length to receive the mantel assembly. I slid the mantel over the ledger and fastened it with countersunk screws and oak plugs, which the painters later stained to match.

I supported the mantel with 2x4s until the pilasters arrived, and then happily finished the job *(Figure B)*.

Figure A. *Mantelpieces can be installed from the top down, if necessary. Be sure the backing is well secured to the wall. I used several long drywall screws into each stud to secure this monster.*

Figure B. *And I propped it up with 2x4s temporarily, until the pilasters came from the finisher and could be installed.*

to the interior dimension of the pilaster box, and subtract 1/16 in. to avoid any problem getting the pilaster over the backing. Fasten the backing board to the wall (nails or screws and adhesive caulk), and then fasten the pilasters to the backing blocks with 8d finish nails, through the sides.

The frieze. Once the pilasters are installed, fasten a backing block to the wall for the hollow frieze, just like the pilasters. Then cut the frieze to fit tightly between the pilasters, and secure it to the backing block with 8d finish nails from the top and bottom *(Figure 15-33)*.

The mantelshelf. On this simple mantelpiece, we went straight for the mantelshelf and assembled the shelf, crown molding, and nosing all in place, right on the wall *(Figures 15-34 and 15-35)*.

Rich molding for decoration. The simplicity, ease of construction, and inexpensive framework of this modest design are enhanced by somewhat expensive moldings.

The fluting is a full 1 1/2 in. thick, and barrel-shaped (in the past, I've combined several pieces of this molding to form a complete column). The oversized rosettes are also more expensive than standard ornaments, and to provide a sufficiently deep point of termination for the fluting at the baseboard, we first attached a back out—an additional piece of MDF backing behind the baseboard *(Figures 15-36 to 15-39)*.

A Brick Surround with Flat Pilasters

There's a big difference between building a mantel on a flat wall and wrapping one around an exist-

Figure 15-36. *Like casing a doorway, the rosettes should be applied first, with glue and nails.*

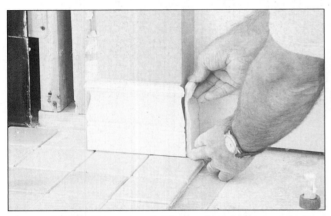

Figure 15-37. *To prevent the fluting from being proud of the base molding (the fluted molding in this example is curved and almost 1¼ in. thick!), add backing behind the base.*

Figure 15-38. *Cut and install the fluting on the pilasters, and then finish the fluting across the frieze.*

Figure 15-39. *As for proportions: Notice that the length of the mantel was limited by the window casing, and the height of the mantel was determined by the fireplace opening.*

ing brick surround. One way to minimize the work involved is to use a flat backboard. Even easier is a flat pilaster, as in this example.

The homeowner on this job wanted me to incorporate the existing brick surround inside the mantelpiece, so that it looked as if the brick had been added later. So we built L-shaped pilasters that wrap around the outside corners of the brick surround *(Figure 15-40)*.

The pilasters. We made the short leg of the L a little longer than necessary, and then scribed it to fit the contour of the wall, while simultaneously allowing the long face of the L to lie flat against the brick *(Figure 15-41)*. Then we fastened the pilasters to the brick using plastic anchors, drywall screws, and adhesive caulk *(Figure 15-42)*.

We scribed the second pilaster to the marble hearth before fastening it to the brick, making sure that the pilasters were level with each other *(Figure 15-43)*.

The frieze and pilaster decoration. The frieze was even easier, just a single piece of MDF which we fastened to the brick with anchors, screws, and adhesive. Then we started on pilaster decorations, because the end blocks would help support the mantelshelf.

Though the basic framework of this mantel is similar to the previous example, economy was the key to this design. We used an easily affordable fluted casing profile to decorate the pilasters, and

Figure 15-40. *We build L-shaped pilasters to wrap around a brick surround.*

Figure 15-41. *This example is the easiest to install and needs to be scribed only to the rear wall, not to the brick.*

Figure 15-43. *Scribe just enough off the bottom so that the pilasters fit tightly around the hearth, but be sure the tops of the pilasters are level.*

Figure 15-42. *Use anchors, screws, and adhesive caulk to fasten the pilaster to the brick.*

Figure 15-44. *Install the base block, and then the pilaster fluting (in this case, an inexpensive profile), then the end blocks.*

flat blocks of MDF for the base and end blocks **(Figure 15-44)**. With the end blocks in place, we had a more stable surface for the mantelshelf.

The mantelshelf. In this case, the shelf had to be scribed to wrap around the brick surround. First we made a rough-cut: We held the shelf on top of the frieze and marked each shoulder cut at the corners of the brick. On the first attempt, we cut right on the shoulder marks but purposely made the cut 1/4 in. shy of the brick-surround face. Then, for the second and final cut, we scribed the shelf to the face of the surround and to the walls at each end. To complete the shelf and cover the seam between the two pieces of MDF, we applied a nosing on the front edge **(Figures 15-45 to 15-47)**.

This design is economical, but because the crown molding breaks forward around the shallow end blocks, the mantelpiece nonetheless has a rich appearance **(Figure 15-48)**. Yes, those crown returns were only 3/4 in. from long point to short

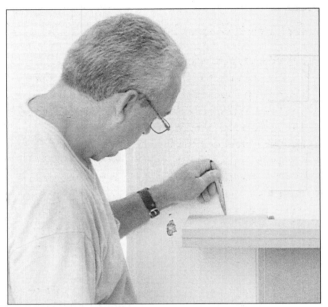

Figure 15-45. *Scribe the mantelshelf to fit snugly around the brick surround and tight against the front and back walls.*

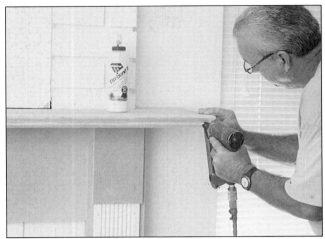

Figure 15-47. *Install the nosing with brad nails and glue.*

Figure 15-46. *Secure the mantelshelf to the back walls with pocket holes and screws, and then nail the shelf down into the pilasters, too.*

Figure 15-48. *Even this simple mantelpiece has a dramatic effect because the crown molding breaks around the pilasters.*

point. Those pieces aren't difficult to cut from a long piece of molding; just be sure to get them right on the first cut.

A Brick Hearth and Deep Pilasters

Laying flat pilasters or backboards directly on the brick is an easy solution to the problem of installing a mantelpiece on an existing brick surround. But the challenge of scribing a full pilaster to a brick surround—and the substantial increase in labor expense—also has a substantial reward: The deeper shadows result in a much more dramatic mantelpiece *(Figure 15-49)*.

The pilasters. Another finish carpenter and I built the mantelpiece in the following photographs for an extremely picky homeowner—his wife, who wanted the mantelpiece to be an eye-catcher.

We started by assembling U-shaped pilasters, with the outside leg of each pilaster long enough to reach around the brick surround and touch the wall. We scribed the first pilaster to fit the hearth first, and then we scribed it to fit the wall and the surround. We followed the same steps for the second pilaster (which I'll demonstrate next), but first we matched the height of the two pilasters, as follows:

My colleague measured the difference in height between the two pilasters with a long level, then spread his scribes that exact amount, and scribed the bottom of the pilaster to fit right into the brick hearth, just as he had the first pilaster *(Figures 15-50* and *15-51)*.

Figure 15-49. *For even more drama, build out the pilasters and increase the projection of the mantelpiece, but don't always break the crown molding around the pilasters!*

Figure 15-50. *Build U-shaped pilasters to increase the depth around a brick surround.*

Figure 15-51. *Use a long level to determine how much to scribe off each pilaster.*

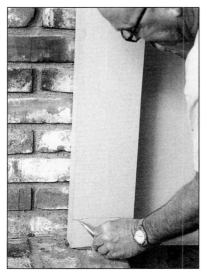

On-the-Wall Pilasters

I don't always preassemble pilasters. In fact, sometimes I start a mantelpiece without a clear idea of where I'm going or what the finished mantel will look like. Besides, if your tools are all set up in front of the fireplace, sometimes it's easier to assemble everything right on the wall.

After laying out the pilasters on the wall for this mantel, I nailed two layers of backing onto the wall (the second layer would be the panel face) and then wrapped the backing with mitered MDF *(see photo)*. The front pieces that wrap around the face of the pilaster are intentionally short, so as to leave room for recessed panel molding.

Pilasters can also be built directly on the wall, especially if your mobile shop is located in the same room. The short mitered returns on the sides of this pilaster will be finished with recessed paneling.

After cutting the pilaster to fit the hearth, we scribed along the length of the pilaster so that it would fit right into the grout joints of the surround. This is a slow and painstaking process, as the scribes must be held perfectly level (perpendicular to the brick) at all times, especially while scribing for the shoulder cuts around each brick *(Figure 15-52)*. In a moment you'll notice that we made a sloppy error in this regard near the bottom of the pilaster, but thankfully this was a paint-grade mantelpiece (of course we filled it before the homeowner saw it!).

Cutting the scribe line is also a painstaking yet rewarding challenge. If parts of the job — the scribing and the cutting — are performed with care, the pilaster will appear to have been installed before the brick surround *(Figures 15-53 and 15-54)*.

The mantelshelf. Because the hearth already projected several inches into the room — and the pilaster legs projected even farther into the room — we decided the mantelshelf didn't need the added depth of breaking crown molding, so we preassembled the entire mantelpiece at the saw.

MANTELPIECES 253

Figure 15-52. Scribe the bottom of each pilaster into the hearth (get it right into the grout joints!).

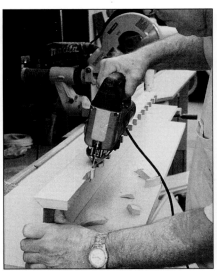

Figure 15-53. Follow the scribe line with a fine-cutting jigsaw blade, and back-cut from the scribe line to ensure a tighter fit.

Figure 15-54. Patience and care will result in a tight fitting scribe on the first try. Poor-fitting scribes (notice both lower grout-joint scribes) are usually the result of not holding the scribes perfectly level and perpendicular to the wall.

We started the mantelshelf by cutting the lower soffit, which was made deep enough and long enough to cantilever 1/2 in. past the end blocks — an equal amount on the ends and the face.

Next, we cut and fastened the crown molding to the soffit, using miter clamps, glue, and brad nails *(Figure 15-55)*.

After completing the soffit and crown, determining the size of the mantelshelf was easy and didn't require any mind-bending math. We simply made the shelf 1/2 in. longer and deeper than the top of the crown molding. And we added another 1/2 in. for the beaded nosing.

I'd forgotten to bring along a piece of molding for the nosing, so we used a bearing-guided router bit to cut a beaded edge on the lower piece of the laminated shelf. We later added a piece of bead-and-barrel molding to the upper piece of the shelf, which completed the nosing detail.

Decorating the pilasters and frieze. I won't go into elaborate detail about the methods we used to decorate the pilasters, frieze, and end blocks. A close look will reveal that we used rabbeted panel molding (left over from a previous job). But rather than making the extra effort to build face frames for recessed panels, we reversed the molding and applied it to additional panels mounted onto the face of the mantelpiece *(Figure 15-56)*. That saved us a lot of time and still provided the owner with the type of separation she required for the painting design she had in mind.

Figure 15-55. Eliminate math whenever possible: Preassemble the soffit and crown molding, then measure the top of the crown molding to determine the size of the mantelshelf.

Figure 15-56. Save the decorations for last. In this example, notice the reversed panel molding and flat panel inserts. Here, the last brad is being nailed into the architrave molding, separating the top of the pilaster from the bottom of the entablature.

Appendix 1

Suppliers, Products, and Resources

Thank you to the many manufacturers who kindly supplied materials, tools, and advice.

SUPPLIERS

Andersen Windows
651/264-5150
www.andersenwindows.com

Bosch Power Tools
877/Bosch99
www.boschtools.com

Delta
800/223-7278
www.deltawoodworking.com

DeWalt
800/4-Dewalt
www.dewalt.com

Drill Doctor
800/597-6170
www.drilldr.com

Eagle Window & Door, Inc.
800/453-3633
www.eaglewindow.com

Enkeboll Designs
800/745-5507
www.enkeboll.com

Far West Plywood
818/885-1511

FastCap
888/443-3748
www.fastcap.com

Festool
Festool Plunge Router
888/337-8600
www.festool-usa.com

Forest Plywood
800/936-7378
www.forestplywood.com

Hitachi
800/829-4752
www.hitachi.com

Jeld-Wen
800/459-6734
www.jeld-wen.com

J.P. Weaver
Composition Molding & Petitsin
818/500-1740
www.jpweaver.com

Maestro Company
888/263-7501

Makita
800/462-5482
www.makitatools.com

Marvin Windows and Doors
800/537-7828
www.marvin.com

Micro-Fence
800/480-6427
www.microfence.com

Milwaukee Electric Tool Corp.
800/729-3878
www.milwaukeetool.com

Occidental Leather
707/824-2560
www.bestbelt.com

Panasonic
800/338-0552
www.panasonic.com

Pella
800/374-4758
www.pella.com

Porter Cable
800/487-8665
www.porter-cable.com

Pozzi Wood Windows
800/922-6222
www.pozzi.com

Ryobi Tools
800/525-2579
www.ryobitools.com

Senco
800/543-4596
www.senco.com

Stabila
800/869-7460
www.stabila.com

Superior Moulding
800/473-1415
www.superiormoulding.com

Templaco Tools
800/578-9677
www.templaco.com

Woodworker's Supply
800/645-9292
www.woodworker.com

PRODUCTS

Boring Jig
Classic Engineering
866/267-3544
www.boringjigs.com

Collins Spring Clamps
Collins Coping Foot
Collins Tool Company
888/838-8988
www.collinstool.com

Construction Master Pro Calculator
Calculated Industries, Inc.
800/854-8075
www.calculated.com

Double-Edge Melamine Trimmer
Virutex
800/868-9663
www.virutex.com

Gizmo Line Laser Level
CST/Berger
815/432-5237
www.cstsurvey.com

Hinge Templates and Boring Jigs
Templaco Tools
800/578-9677
www.templaco.com

Jiffy Connectors
Häfele America
800/423-3531
www.hafeleonline.com/usa

Kreg Jig K2000 ProPack
Kreg Tool Company
800/447-8638
www.kregtool.com

Metal Jambs
Timely Industries
800/247-6242
www.timelyframes.com

Moistop & E-Z Seal
Fortifiber Building Systems Group
800/773-4777
www.fortifiber.com

Paslode Angled Finish Nailer
847/634-1900
www.paslode.com

PLS-5 Laser System
Pacific Laser Systems
800/601-4500
www.plslaser.com

Rol-Air Compressors
Rol-Air Systems
920/349-3281
www.rolair.net

Rotozip Spiral Saw
Bosch Tool Corp.
877/ROTOZIP
www.rotozip.com

Rousseau Saw Stand
Rousseau Company
800/635-3416
www.rousseauco.com

Sawhelper Ultrafence
American Design and Engineering
800/441-1388
www.sawhelper.com

Scribes
General Tool Company
800/314-9817
www.gentool.com

Stiletto Titanium Hammers
Stiletto Tools
800/987-1849
www.stilettotools.com

Trojan Saw Stand
Trojan Manufacturing, Inc.
800/745-2120
www.trojantools.com

Ulmia Spring Clamps
Garrett Wade Tool Catalog
800/221-2942
www.garrettwade.com

Weatherstripping
Pemko Manufacturing
800/283-9988 (West Coast)
800/824-3018 (East Coast)
www.pemko.com

Zircon Stud Finder
Tool Crib
800/635-5140
www.toolcrib.amazon.com

RESOURCES

Bevel and angle-miter charts
(for cutting crown moldings)

www.dewalt.com/us/woodworking/articles/article.asp
www.issi1.com/corwin/crown.html
www.josephfusco.org

APPENDIX 2

Takeoff Charts

Exterior Door Takeoff

Date: _____ Job: _____
Location: _____
Manufacturer: _____ Type: _____

Item #	Quantity	Dimensions	Jamb Width	Exterior Trim	Hand	Sill	Special Hardware	Plan#	Description/ Location	Glass Type	Lites	R/O Size
#1												
#2												
#3												
#4												
#5												
#6												
#7												
#8												
#9												
#10												

Interior Door Takeoff

Date: _____ Job: _____
Location: _____
Manufacturer: _____ Type: _____

Item #	Quantity	Dimensions	Jamb Width	Hand	Stop Size and Style	Hardware Type	Plan #	Description/ Location	Door Type	Undercut	R/O Size
#1											
#2											
#3											
#4											
#5											
#6											
#7											
#8											
#9											
#10											

Window Takeoff

Date: _____ Job: _____

Location: _____

Manufacturer: _____ Type: _____

Item #	Quantity	Dimensions	Jamb Width	Plan #	Description/ Location	Window Type and Hand	Glass Type	Lites	R/O Size	Cost Each	Sub-total Material	Labor Each	Sub-total Labor
#1													
#2													
#3													
#4													
#5													
#6													
#7													
#8													
#9													
#10													

Appendix 3

Crown Molding Miter & Bevel Angle Setting Chart

Crown Molding Miter & Bevel Angle Setting Chart

Wall Angle	38/52 miter	38/52 bevel	45/45 miter	45/45 bevel
1	89.18	51.99	89.29	44.99
2	88.37	51.98	88.58	44.99
3	87.56	51.97	87.87	44.98
4	86.75	51.95	87.17	44.96
5	85.94	51.93	86.46	44.94
6	85.13	51.89	85.76	44.92
7	84.32	51.86	85.05	44.89
8	83.52	51.82	84.35	44.86
9	82.71	51.77	83.64	44.82
10	81.91	51.72	82.94	44.78
11	81.11	51.66	82.24	44.73
12	80.31	51.60	81.54	44.68
13	79.51	51.53	80.84	44.63
14	78.72	51.45	80.14	44.57
15	77.92	51.37	79.45	44.51
16	77.14	51.29	78.75	44.44
17	76.35	51.20	78.06	44.37
18	75.57	51.10	77.37	44.29
19	74.79	51.00	76.68	44.21
20	74.02	50.89	75.99	44.13
21	73.24	50.78	75.31	44.04
22	72.47	50.67	74.62	43.95
23	71.71	50.55	73.94	43.86
24	70.95	50.42	73.26	43.76
25	70.19	50.29	72.59	43.65
26	69.44	50.15	71.91	43.54
27	68.69	50.01	71.24	43.43
28	67.95	49.87	70.57	43.32
29	67.21	49.72	69.91	43.20
30	66.48	49.56	69.24	43.07
31	65.75	49.40	68.58	42.95
32	65.02	49.24	67.92	42.82
33	64.30	49.07	67.27	42.68
34	63.59	48.90	66.61	42.54
35	62.88	48.72	65.96	42.40
36	62.17	48.54	65.32	42.26
37	61.47	48.35	64.67	42.11
38	60.78	48.16	64.03	41.95
39	60.09	47.97	63.39	41.80
40	59.40	47.77	62.76	41.64
41	58.73	47.57	62.13	41.47
42	58.05	47.36	61.50	41.31
43	57.38	47.15	60.87	41.14
44	56.72	46.93	60.25	40.96

Wall Angle	38/52 miter	38/52 bevel	45/45 miter	45/45 bevel
45	56.06	46.72	59.63	40.78
46	55.41	46.49	59.02	40.60
47	54.76	46.27	58.41	40.42
48	54.12	46.04	57.80	40.23
49	53.49	45.81	57.19	40.04
50	52.85	45.57	56.59	39.85
51	52.23	45.33	55.99	39.65
52	51.61	45.09	55.40	39.46
53	50.99	44.84	54.81	39.25
54	50.38	44.59	54.22	39.05
55	49.78	44.34	53.64	38.84
56	49.18	44.08	53.05	38.63
57	48.59	43.83	52.48	38.41
58	48.00	43.56	51.90	38.20
59	47.42	43.30	51.33	37.98
60	46.84	43.03	50.76	37.76
61	46.26	42.76	50.20	37.53
62	45.69	42.48	49.64	37.30
63	45.13	42.21	49.08	37.07
64	44.57	41.93	48.53	36.84
65	44.0	41.65	47.98	36.61
66	43.47	41.36	47.43	36.37
67	42.92	41.08	46.89	36.13
68	42.38	40.79	46.35	35.88
69	41.85	40.49	45.81	35.64
70	41.32	40.20	45.28	35.39
71	40.79	39.90	44.75	35.14
72	40.27	39.60	44.22	34.89
73	39.76	39.30	43.69	34.63
74	39.24	39.00	43.17	34.38
75	38.74	38.69	42.66	34.12
76	38.23	38.38	42.14	33.86
77	37.74	38.07	41.63	33.59
78	37.24	37.76	41.12	33.33
79	36.75	37.44	40.62	33.06
80	36.26	37.13	40.12	32.79
81	35.78	36.81	39.62	32.52
82	35.30	36.49	39.12	32.25
83	34.83	36.17	38.63	31.97
84	34.36	35.84	38.14	31.70
85	33.89	35.51	37.65	31.42
86	33.43	35.19	37.17	31.14
87	32.97	34.86	36.69	30.85
88	32.51	34.53	36.21	30.57
89	32.06	34.19	35.73	30.28
90	31.61	33.86	35.26	30.00

Wall Angle	38/52 miter	38/52 bevel	45/45 miter	45/45 bevel
91	31.17	33.52	34.79	29.71
92	30.73	33.19	34.32	29.41
93	30.29	32.85	33.86	29.12
94	29.86	32.50	33.40	28.83
95	29.43	32.16	32.94	28.53
96	29.00	31.82	32.48	28.23
97	28.57	31.47	32.03	27.93
98	28.15	31.13	31.57	27.63
99	27.73	30.78	31.12	27.33
100	27.32	30.43	30.68	27.03
101	26.90	30.08	30.23	26.73
102	26.49	29.73	29.79	26.42
103	26.09	29.37	29.35	26.11
104	25.68	29.02	28.92	25.80
105	25.28	28.66	28.48	25.49
106	24.88	28.31	28.05	25.18
107	24.49	27.95	27.62	24.87
108	24.09	27.59	27.19	24.55
109	23.70	27.23	26.76	24.24
110	23.32	26.87	26.34	23.92
111	22.93	26.50	25.91	23.61
112	22.55	26.14	25.49	23.29
113	22.17	25.78	25.08	22.97
114	21.79	25.41	24.66	22.65
115	21.41	25.05	24.25	22.33
116	21.04	24.68	23.83	22.00
117	20.6	24.31	23.42	21.68
118	20.30	23.94	23.02	21.35
119	19.93	23.57	22.61	21.03
120	19.56	23.20	22.20	20.70
121	19.20	22.83	21.80	20.37
122	18.84	22.45	21.40	20.04
123	18.48	22.08	21.00	19.71
124	18.12	21.71	20.60	19.38
125	17.77	21.33	20.20	19.05
126	17.41	20.96	19.81	18.72
127	17.06	20.58	19.42	18.39
128	16.71	20.20	19.02	18.05
129	16.36	19.83	18.63	17.72
130	16.01	19.45	18.24	17.38
131	15.67	19.07	17.86	17.05
132	15.32	18.69	17.47	16.71
133	14.98	18.31	17.09	16.37
134	14.64	17.93	16.70	16.03
135	14.30	17.55	16.32	15.69

This chart is reprinted with permission from carpenter and cabinetmaker Joe Fusco's website. To make your own copy, visit www.josephfusco.com.

Index

A

Adjustable shelving
    for bookshelves, 222, 223
    for closets, 133–134
    See also Shelving.

Adjusting
    door to fit jamb (hinges), 65–68
    double-door frames, 56–57
    jambs for new doors, 90–91
    sliding doors, 45, 53–54
    sliding windows, 73

Apprenticeship, introduction to training, 1

Apron, installing, 110

Arches
    arched casings, 203-212
    arched jambs, 197–202
    layout of radius, 206-212

Architecture
    effect on finish carpentry, 2
    recommended reading, 2
    See also Classical orders.

Arts-and-Crafts style
    bookcase, 219–228
    double-hung window layout, 70
    overview, 8–9

B

Baltic birch (multi-ply), 21

Baseboard, 115–124
    base cap and base shoe, 17–19
    built-up, 121
    coped joints, 118–120
    curved walls, 123–124
    cutting double outside corners, 117
    cutting inside corners, 116–117
    developing cut list for, 115–116
    installing efficiently, 120–121
    measuring properly for, 116
    round corners in, 19, 122
    self returns, 116, 118
    stain-grade radius, 123–124
    variety of styles, 17–19

Beadboard, 177

Benches
    portable door, 63
    tool, 39–40, 41

Biscuit joinery, versus splines, 166, 203

Bookcases, 219–236
    construction details, 219–222, 229
    construction of, 222–224
    installation, 225–232, 235-236
    suspended, 232–236

Bosch tools
    1587VS jigsaw, 37
    1613 EVSB router, 35
    1632VS reciprocating saw, 37
    Angle Finder, 145
    drills, 36

Bosses, 192

Bullnose corner
    characteristics, 19
    installing baseboard, 122
    not for crown, 151

Butt-in-the-butt method, 44

C

Calculator
    for laying out arched doorways, 199, 209-210
    for laying out wainscoting, 160–161

Carved moldings, characteristics of, 14–15

Casement windows
    building new jambs for, 75
    weatherstripping for, 75–76

Casing, 99–114
    arched, 203-212
    correcting out-of-plumb window by reinstalling, 229
    cutting miters, 99–100
    for arched doorways, 203-212
    for metal jambs, 97-98
    hollow-core prehung doors before installation, 80
    installing door, 100–101
    measuring door, 100
    multiple windows, 111–113
    scribing, 101–104
    variety of styles, 16–17
    with keystones, 208
    with rosettes and plinth blocks, 203–207

Caulk
    comparing urethane and acrylic latex, 49
    for baseboard, 121, 122
    for closet shelving, 135
    for coffered ceilings, 189
    for columns, 215–218
    for crown molding, 150–153
    for doors and windows, 49
    for exterior door frames, 49
    for exterior door sills, 46
    for jamb shims on concrete floors, 84
    for mantelpiece pilasters, 248-249, 250
    for over-door trim, 214
    for soffit moldings, 187
    for window installation, 71, 74
    miscellaneous supplies, 42
    with door flashing, 50

Ceilings, decorative, 181–195
    bosses, rosettes, and pendants for, 192
    coffered, 188–195
    installing stiles and rails, 182
    layout, 181–182
    light wells and soffits, 185
    odd-shaped panels, 182–185
    paneled, 181–185
    soffits, 185–186
    solving problems using new design, 195–196

Chair rail
    establishing height for plant-on paneling, 159
    overview, 20
Circular saws, 33
Classical orders
    definition, 2
    five, 2–4
    in American architectural styles, 4–11
    in decorative doorway design, 197
    in mantelpiece design, 237–241, 245
    in soffit moldings, 187
    in wainscoting, 159
    need for familiarity with, 1–2
    relation to trim details, 4
    See also Design and layout.
Closet shelving, 125–139
    accessories, 129–130
    adjustable, 133–137
    installing paint-grade, 130–132
    installing prefinished, 132–133
    iron-on edge tape for, 136
    paint-grade versus prefinished, 128–129
    shapes and sizes, 125–128
Collins Coping Foot®, 37, 120, 148
Columns
    classical design, 1–5
    in decorative doorways, 215–218
    in Federal style, 6–7
    in Georgian style, 5–6
    in Modern style, 10
    in Victorian style, 7–8
Compressors, 32–33
Construction Master Pro®. *See* Calculator.
Coped joinery
    cutting in baseboard, 119–120
    cutting in crown molding, 147–148
    measuring in baseboard, 118–119
    versus mitering, 118
    with Collins Coping Foot®, 37, 120, 148
Cordless tools, 35–36
Cornices
    Georgian-style, 6
    in classical orders, 3
    on bookcases, 231–232
    on mantelpieces, 237, 239, 246–247
    on over-door trim, 213-214
Cross-legged jambs, 51–52, 83–84
Crown molding, 141–158
    coping, 147–148
    cutting acute angles in, 156–158
    cutting miters in, 142–148
    finding angles for, 145–147
    finding studs for, 148
    for bookcases, 226-228
    in coffered ceilings, 191-195
    installing, 148–155
    layout and measurement for, 141
    overview and variety of styles in, 19
    patching, 155–156
    preassembling splices in, 150
    with full-height wall paneling, 175

D

Decorative ceilings. *See* Ceilings, decorative.
Decorative doorways. *See* Doorways, decorative.
Decorative walls. *See* Walls, decorative.
Design and layout
    in bookshelves, 219–222, 228-230, 232–234
    in built-up baseboard, 121
    in built-up casing, 212-213
    in chair rail, 177–178
    in closets, 125–128, 133, 134
    in coffered ceilings, 188-189
    in decorative ceilings, 181–182
    in decorative doorways, 197, 213
    in diagonal paneling, 178–179
    in full-length wall paneling, 174–177
    in mantelpieces, 237–247, 249–250, 252–254
    in multiple-window casing, 111–113
    in problem ceilings, 195–196
    in soffit moldings, 185–188
    in wainscoting, 159–165
    need for familiarity with heritage
        of styles, 1–2, 5
    See also Classical orders.
Doors, 43–68, 79–98
    casing for, 100–104
    flashing exterior, 57–58
    handing, 44
    hinge templates, 91
    homemade hook for, 60
    installing exterior frames, 45–53
    installing hollow-core prehung, 79–93
    installing metal jambs, 94–95
    installing new doors in new jambs, 88–93
    installing new doors in old jambs, 58–68
    installing prefit, 95–98
    installing sliding and swinging, 53–57
    installing solid-core prehungs and pairs, 84–88
    ordering exterior, 43–45
    ordering interior, 79
    ordering metal jambs, 93–94
    portable bench for, 63
    sill pan for, 47
    undercut and beveled (UB), 88
Doorways, decorative, 197–217
    arched, 197–202
    columns and pilaster for, 215–218
    construction approaches for
        arched jambs, 197–198
    Federal style, 6–7
    Georgian style, 5–6
    installing arched jambs, 199–202
    over-door trim, 213–214
Drills
    overview, 36

using two for installing window extensions, 109–110

E
Embossed moldings, 15
Extension cords, choosing, 32
Extension jambs. *See* Jamb extensions.

F
Federal style
    in mantelpiece design, 239–241
    overview, 6–7
    window-pane layout, 70
Finger-jointed moldings, 14
Flashing
    exterior door jambs, 50
    windows, 71, 74
Flat- (or plain-) sliced veneer, 22–23
Flexible moldings (flex-trim)
    characteristics, 14
    for paint-grade radius baseboard, 123
    for radius doorway casing, 203–206
    in ceiling corners, 187
    in moldings for curved panels, 174
Flexible plywood, 23–24
Fluting
    on mantelpiece pilasters, 240-242, 243
    on radius casing, 210–212
Friezes
    in classical design, 3, 4
    in Elizabethan-Gothic style, 5
    in Georgian style, 6
    on mantelpieces, 242–246, 249, 250-251, 254
Fypon®, 14, 239

G
Georgian style
    in decorative doorway design, 197, 213
    in mantelpiece design, 239-240, 243-244
    overview, 5–6
    window-pane layout, 70

H
Handing
    doors at a door bench, 89
    for doors, 44
    for windows, 69
Hinge-side method, 44
Hinges
    adjusting, 65-66
    for casement windows, 75
    full-length templates for, 91
    installing on new doors in new jambs, 89–90
    locating, 60
    marking location with tape, 88
    mortising for on exterior doors, 62
Hitachi C15FB sliding saw, 26

I
Inlay, decorative, 225
Installation
    adjustable shelving, 133–137
    arched doorways, 199–202
    baseboard, 120–121
    bookshelves, 225–228, 230–232, 235–236
    ceiling stiles and rails, 182
    chair rail, 177–178
    coffered ceilings, 188–195
    columns, 215–218
    crown, 148–155
    diagonal paneling, 178–179
    door casing, 100–101
    doorway rosettes and plinth blocks, 203–207
    exterior door frames, 45–53
    hollow-core prehung doors, 79–84
    lockset, 62–68, 90
    mantelpieces, 238–254
    metal jambs and prefit doors, 94–98
    new doors in new jambs, 88–93
    new doors in old jambs, 58–68
    over-door trim, 214
    paint-grade closet shelving, 130–132
    prefinished closet shelving, 132–133
    radius baseboard, 123–124
    round corners, 122
    sliding door locks and hardware, 55
    sliding doors, 53–54
    solid-core prehungs and pairs, 84–88
    solid-sheet and T&G paneling, 177
    swinging doors, 54–57
    wainscoting molding, 164, 170–173
    wall-paneling stiles and rails, 168
    wall panels, 173
Iron-on edge tape
    applying to shelving components, 136
    for arched doorway jambs, 202

J
Jamb extensions
    for stool-and-apron windows, 107–110
    manufacturer's, 104–105
    milling, 105–106
    narrow, 102
    picture-frame, 106
    scribing for irregular walls, 106–107
    site-built for windows, 105
    stacking, 106
Jigsaws
    Collins Coping Foot® for, 37, 120, 148
    for countertop scribes, 103–104
    for cutting chair-rail scribes, 177–178
    for panel cutouts, 170
    for paneling scribes, 178
    overview, 37

K

Kortron®
for prefinished shelving, 128–129
overview, 24

Kreg Mini Jig®, 230, 231, 235

L

Laser levels
for control lines, 52
for layout, 39

Layout. *See* Design and layout.

Levels
in squaring door frames, 51
marking hinge locations with tape, 88
Master spring-loaded, 85
overview, 38–39
setting prehungs out-of-plumb, 81
setting windows, 71–72
shooting control lines, 52
Stabila magnetic, 85
use for bookshelf toe kicks, 221
use in building arched door jambs, 201
use in installing metal jambs, 96
use in laying out chair rail and
 wall paneling, 159, 164
use on floor of interior door opening, 84
use on rough sill, 46

Lockset
dimensional standards for, 60-61, 79
installing strike plate, 67-68
locating, 60-61
lock-boring jig, 64
ordering, 44, 93–94
production work, 79
routing for, 63-64, 90
sliding-door, 55

M

Makita 2703 table saw, 29–30
Makita LS1211 sliding saw, 25–26, 143
Mantelpieces, 237–254
Arts-and-Crafts style, 9
Beaux-Arts style, 10
Federal style, 6–7
Georgian style, 6
Modern style, 10–11
naive designs, 1
Victorian style, 8

Maple ply (apple-ply), 21

MDF (medium-density fiberboard)
characteristics as molding, 13–14
characteristics as sheet material, 21–22
difficulty cutting cope joints in, 119–120, 147
for arched doorways, 198-199, 201–202
for built-up baseboard, 121
for closet dividers, 131, 133, 134
for coffered ceiling beams, 190
for crown molding shims, 149–151
for keystones, 208
for mantelpiece components, 239, 242–243, 247–249, 250–251
for mull battens, 113
for paint-grade shelving, 128
not for jamb extensions, 105
versus particleboard, 128

Melamine®
dividers for adjustable shelving, 133–134
for prefinished shelving, 128–129
overview, 24

Milwaukee 6543-1 screw shooter, 36

Miter saws
acute angle fence for, 114
advantages of double-bevel machines, 144
overview, 25–26
repetitive-stop system for, 27
stands for, 26, 28–29, 116
translating scale to true angles, 113-114

Miters
baseboard double outside corners, 117
baseboard inside corners, 116–117
cutting acute angles in crown, 156–158
cutting crown molding in position, 142–143
cutting crown molding on-the-flat, 143–144
cutting panel molding, 163, 165, 170–171
cutting trapped, 112
four angles for crown, 145–147
in diagonal paneling, 178–179
measuring odd angles, 113
on casing, 99–100
on radius casing, 206
self returns, 116, 118
spring clamps for, 152, 173
using Bosch Angle Finder, 145
versus copes, 118

Modern style, 9–10

Moistop®
use in door installation, 50, 58
use in window installation, 71, 74

Mortising
door manufacturer's, 94
doors, 62–65, 68, 90

N

Nail guns
overview, 30–31
repair kits for, 31

Neoclassical style. *See* Federal style.

O

Order
carpenters' contribution to, 2
See also Classical orders.

Ordering building components
exterior doors, 43–45

metal jambs, 93
prehung interior doors, 79
windows, 69–71
Ornaments, of compo and urethane, 15–16
Overmantel, Georgian style, 6

P
Paint-grade material, 13–16, 21. *See also* MDF and Flexible moldings (flex-trim).
Palladian jambs, 199–202
Palladio, on classical proportions, 3–4
Panel saw, high-rpm
for cutting casing to scribe lines, 103
for cutting stiles to scribe lines, 168
for cutting window stools, 108
for fitting replacement windows, 78
Paneling. *See* Walls, decorative and Ceilings, decorative.
Particleboard, versus MDF, 21, 128
Pattern books, 2–3
Pendants, 192
Pilasters
in classical orders, 3
in Federal versus Georgian style, 6
on bookcases, 220
on mantelpieces, 238–246, 248–251, 252–254
Planing
bevel on narrow sash, 77
doors to size, 61–62
Plinth blocks
Federal style, 7
in classical orders, 3
on bookcase, 224–225
on decorative doorways, 207
Plumbing
jamb legs, 52–53
setting prehung doors out-of-plumb, 81
shimming exterior door trimmers, 46–47
solid-core prehungs, 84–86
Plunge routers. *See* Routers.
Plywood, 20–21
Porter-Cable 97529 router, 35
Preassembling
beam-sides for coffered ceilings, 190–191
cornices for mantelpieces, 239
crown for bookcases, 226-228, 232
crown for coffered ceilings, 191
crown molding, 153–154
entire mantelpiece, 253–254
not for urethane moldings, 155
odd-shaped ceiling panels, 184–185, 185–186
over-door trim, 213-214
picture-frame extensions, 106
pilasters for mantelpieces, 238–242
soffit and crown for mantelpieces, 254
splices for crown, 150
stiles and rails for wainscoting, 167
stool-and-apron for windows, 107–108

three-piece crown for round corners, 151
versus assembly in-place for mantelpieces, 248–249, 253-254
wall-panel molding, 172

Q
Quarter-sliced veneer, 23

R
Reciprocating saws, 37
Repetitive-stop system. *See* Miter saws.
Replacement windows, 76–78
Rift-cut veneer, 23
Rosettes
Federal style, 7
on arched casing, 203–207
on coffered ceilings, 192
on mantelpieces, 250
on rectangular jambs, 207
Rotary-cut veneer, 22
RotoZip, for paneling cutouts, 170
Routers
bits for, 35
overview, 34–35
use with Micro Fence®, 210
use with trammel arm, 210–212
Routing
beading for nosing detail, 254
dadoes for bookshelves, 223
for arched doorways, 206–212
for casement hinges, 75
for door latch, 63-64
for exterior door hinges, 62–64
for new interior doors, 90–91
mortise for lock strike, 68
radius flutes, 210–212

S
Sanders, 37–38
Screw shooters. *See* Drills.
Scribing
around brick for mantelpieces, 252–254
bookshelf stiles, 221
casing, 101
chair rail, 177–178
closet shelving, 137–139
jamb extensions, 106–107
new door to old jamb, 58–59
wall paneling, 177, 178
Senco nail guns, 30–31
Sheet goods, 20–24
Shelving
adjustable for bookcases, 222–223
adjustable for closets, 133–134
bookcases, 219–236
dado for bookcases, 221
installing in closets, 135–137
iron-on edge tape, 136

Kortron®, 128–129
MDF, 128
melamine, 128–129
paint-grade for closets, 130–132
prefinished for closets, 132–133
scribing, 137–139
suspended bookshelves, 232–236
See also Bookcases, Closet shelving.

Shimming
behind lock strike, 88
bookshelf toekicks, 221
crown molding, 149-151
exterior door trimmers, 46–47
making perfect shims, 86
not for hollow-core doors, 80
solid-core prehungs and pairs, 84, 85–86
swinging doors, 55–57
UB doors, 89
vinyl windows, 73
windows, 71–72

Sills and waterproofing
dam for casement windows, 75–76
pan for exterior doors, 47–48
pan for windows, 71

Sliding compound miter saws (SCM saws)
for cutting crown, 143–148
overview, 25–26
See also Miter saws.

Sliding doors, installing, 53–54

Spline joints
for ceiling-panel frames, 182
for wall panel frames, 166–169
slot cutters versus biscuit joiners, 203

Stain-grade wood
as material for molding, 13
in moldings for curved panels, 174
in radius baseboard, 123–124
in radius casing, 203, 206–212

Stool-and-apron, installing, 107–110

Story pole
for bookshelves, 222, 234-235
layout template for wainscoting, 161–163

String
to check cross-legged jambs, 51–52
to check straightness of sills, 51, 73

Supplies, miscellaneous, 42

Swinging doors
adjusting double-door frames, 56–57
installing, 54–57

T
Table saws
overview, 26, 29–30
roller stand for, 31
stands for, 30

Takeoff form
for exterior doors, Appendix 2
for interior doors, Appendix 2
for windows, Appendix 2

Timberflex®, 23–24
Tool belt, tools for, 39
Tools, transporting, 40
Trimmers
shimming plumb for exterior doors, 46–47
shimming plumb for interior doors, 84–85

U
Urethane decorative components
characteristics of molding, 14
column capitals, 217-218
crown, 154–155
glue for, 217
ornamentals, 15–16
radius casing, 203

V
V-bronze weatherstripping, 75
Veneer types, 22–23
Victorian style, 7–8

W
Wainscoting. *See* Walls, decorative.
Walls, decorative, 159–179
cutting molding for, 163
diagonal paneling, 178–179
full-height library paneling, 174–177
layout, 159–163, 164, 174–176
molding styles for, 19–20
paneling on radiused, 173–174
plant-on paneling, 159–163
recessed paneling, 164–173

Waterproofing
exterior door sill pan, 47–48
materials, 48
weatherstripping for casement windows, 75–76
windows, 71

Windows, 69–78
casing for, 104–114
flashing, 71
Georgian-style Palladian, 7
hanging casement sash, 74–76
installing, 71–74
jamb extensions, 105–107
ordering, 69–71
replacing double-hung, 76–78
shimming vinyl, 73
sill pan for, 71, 75–76
stool-and-apron, 107–110
waterproofing, 71
weatherstripping, 75–76

Notes

Notes